PRAISE FOR *IKE'S ROAD TRIP*

"*Ike's Road Trip* is a great ride in every sense. Combining deep research with stylistic verve, Brian C. Black takes us back to a formative moment in the American Century, when a young Dwight D. Eisenhower led a convoy of military vehicles across the United States. The experience transformed Eisenhower, and would ultimately transform America as well, with ramifications for our current moment. This is history at its most engaging."
—Ted Widmer, author of *Lincoln on the Verge*

"An epic story—and a reminder that we desperately need twenty-first century visionaries who will do as much to put us *off* the hydrocarbon road."
—Bill McKibben, author *The Flag, the Cross, and the Station Wagon*

"Although the energy transition began before the Interstate Highway System was initiated, Ike understood from the transcontinental convoy of 1919 and during the fight for the German Autobahn during WWII that no modern society could exist without the capacity to link itself into one cohesive country."
—Susan Eisenhower, author of *How Ike Led*

"No president did more to cement America's attachment to driving than Eisenhower, and nothing did more to convince Ike of the value of a national highway system than his 1919 military convoy from New York to San Francisco. Brian C. Black tells the story of that journey in the context of American energy, transport, economic, and military history in crisp and convincing prose. Deploying a talent shared with Eisenhower, Black recognizes the links between the small details and the larger picture—none larger than the history of energy transitions."

—J.R. McNeill, author of *The Webs of Humankind*

"Everyone loves a road trip, and Brian C. Black's vivid account of Dwight Eisenhower's convoy across America in 1919 is both fun and enlightening. Ike's arduous, adventure-filled trek ultimately inspired a revolution in the way we live."

—*Adam Rome*, author of *The Bulldozer in the Countryside*

"Getting an army convoy across the United States in 1919 through dust and dirt, mud and non-existing roads was no small feat. Getting the United States to adopt a transportation system based on ever-growing usage of petroleum was a monumental shift with far-reaching consequences. In this fast-paced and well-told book, Brian C. Black illuminates how the convoy and the larger story are intertwined and what all of this means today, in the age of energy transitions. *Ike's Road Trip* is an accessible and insightful book whose issues resonate today more than ever."

—Thomas Zeller, author of *Consuming Landscapes*

"*Ike's Road Trip* is an insightful and enjoyable take on America's long love affair with cars and roads. Black deftly guides readers through the 1919 convoy's influence on a young Dwight Eisenhower and its role in inspiring Ike's 1956 interstate highway program. In clear and conversational writing, Black illuminates an important tenet of energy history—how fossil fuel use has had profound impacts on American life and culture."

—Raechel Lutz, co-editor of *American Energy Cinema*

"Brian C. Black's account of a special moment in Dwight Eisenhower's storied life—the 1919 cross country military convoy—is an eye-opener. That experience foreshadowed an energy transition premised on unlimited access to fossil fuels and Eisenhower's future advocacy of a national highway system. Because Black salts his narrative with reminders of our contemporary energy transition, *Ike's Road Trip* enriches Eisenhower historiography *and* encourages readers to ponder energy choices they will face."

—Michael J. Birkner, editor of *Democracy's Shield*

"Brian C. Black's wonderful telling of *Ike's Road Trip* introduces readers to a little-known story about an American icon of the twentieth century. It does so within a fascinating context of breathtaking changes in technology, changes in American patterns of travel, and reminds readers of the human stories about the cars we drive, the oil that feeds them, and the roads we travel.

—Edward T. Linenthal, author of *Sacred Ground*

"Brian C. Black takes his readers on a thrilling ride through the life of Dwight D. Eisenhower, along the greatest road-building endeavor of all time, and into American petro-modernity."

—Tyler Priest, author of *The Offshore Imperative*

"Most accounts of Eisenhower's vision of the U.S. Interstate Highway System begin with his experience of the Reichsautobahn during World War II, but *Ike's Road Trip* takes us back to 1919 and an earlier, but equally formative experience for the future five-star general. That summer, as a not quite thirty-year old lieutenant colonel, Ike spent sixty-two days as a member of the Army's First Transcontinental Motor Convoy, a company of nearly three hundred enlisted men and officers who piloted a fleet of trucks, cars, and motorcycles across 3200 arduous miles. Following in the wheel ruts of such automobile pioneers as Horatio Nelson Jackson (1903) and Alice Huyler Ramsey (1909), Ike's convoy was not the first to make the coast-to-coast journey, but no previous trip had mustered comparable promotional ballyhoo and continuous front-page coverage. Nor was there a cross-country excursion that would prove so consequential when, thirty-seven years later, President Eisenhower signed the Federal-Aid Highway Act of 1956 into law. Part picaresque, part military history, *Ike's Road Trip* is also a cautionary tale about the origins of our oil and auto dependency and their twenty-first century consequences."

—Gabrielle Esperdy, author of *American Autopia*

In the summer of 1919, a young Dwight D. Eisenhower departed on one of the most consequential cross-country journeys in modern U.S. history. As Brian C. Black brilliantly shows us, the First Transcontinental Motor Train symbolized the coming of the motor age, highlighted the depressing condition of the nation's roadways, and served as a harbinger of one of the great achievements of Eisenhower's later presidency: the Interstate Highway System. *Ike's Road Trip* is a timely meditation on a monumental energy transition whose consequences remain very much with us today.

—Paul S. Sutter, author of *Driven Wild*

IKE'S ROAD TRIP

IKE'S ROAD TRIP

HOW EISENHOWER'S 1919 CONVOY

PAVED THE WAY

FOR THE ROADS WE TRAVEL

Brian C. Black

GODINE
Boston · 2024

Published in 2024 by GODINE

Boston, Massachusetts

ISBN 9781567927153 (hardcover) | ISBN 9781567927160 (ebook)

Library of Congress Control Number: 2024939501

First Printing, 2024
Printed in the United States of America

To Owen

CONTENTS

Young Eisenhower made his choice and
entered military service proudly.

"It Is Your Choice" [1]

When David and Ida Eisenhower moved their growing family to Abilene, Kansas, in 1898, the town seemed a million miles from a swiftly modernizing world. The cow town's oddly broad main thoroughfare, edged in wood planks, had been made for driving herds of cattle to the rail yard. Beyond, undulating prairie swells bathed every direction in piercing greens. The grassy plains formed a protective cocoon around those growing up there at the time, making life seem less like a moment on the cusp of the twentieth century and more like an earlier era.

"I have found out later that we were very poor," recalled Dwight, the third of David and Ida's seven sons. "But we didn't know it at the time."[2]

Dwight's observation grew from the incredible similarity of the people living around him. With even a modest bit of personal capital, residents of Abilene could introduce their fellow residents to new enterprises and then operate them largely free of competition. With a little grit, success begot success. The Eisenhowers arrived in America in 1741—as Eisenhauer, German for "iron hewer"—having led thriving agriculture efforts in Germany. They continued farming in Pennsylvania, eventually pushing westward to Kansas. David and Ida lived for a short time in Denison, Texas, where Dwight was born, before joining David's father, Jacob Eisenhower, in Abilene. Jacob owned a successful 160-acre farm with a large house that doubled as the Sunday meeting place for the area's Mennonite-like River Brethren. On his property he raised a windmill and a large Dutch barn for his dairy herd, reminiscent of barns he had known in Pennsylvania. With little competition, Jacob bought additional land, founded a local creamery for his milk products, and established a bank in the nearby town of Hope. Jacob promised each of his four children that when they married, he would gift each a quarter section of tillable land and $2,000 in cash; he was a father determined to keep his family close.

The farming life, however, was not David's calling.

Unlike his siblings, David had no interest in agriculture. Instead, with Jacob's support, in 1883 David began studying engineering at nearby Lane College. There he met twenty-one-year-old Ida Stover, who was studying music. Ida had been raised by her maternal grandparents when her parents had died young. Like the Eisenhowers, the Stovers had emigrated from Germany to Pennsylvania before moving on to Virginia. David and Ida were married in 1885. Soon after, the young couple ended their studies and, with wedding funds from Jacob, opened a general store in Hope.

A complete novice to the world of business, David had the sense to partner with Milton Good in the local clothing store. He and Ida took up residence in the small apartment above the shop. After three frustrating months marked by disagreements with his new partner, David borrowed $3,500 from his father and bought out Good. David's younger brother Abraham joined him, and the business was renamed Eisenhower Brothers. As Abraham became increasingly successful in Good's old role, David's interest in the day-to-day operations of the shop waned. In 1888, he walked away from not only the store but Ida and their two sons—both born in Kansas—and moved alone to Denison, Texas. A year later, Ida and the boys joined David in Texas. In 1890, David Dwight Eisenhower, named for his father, was born.

Later biographers of the Eisenhowers report that the family protected David's reputation, and that Dwight died believing that this early move was due to Good's treachery rather than the truth: David's mental instability.[3]

Regardless of the reasons for the move, the result of the uprooting was a financial insecurity that, while foreign to David and Ida, was not uncommon for many Americans living in frontier fringes. For the young family, rock bottom was a one-room shack next to the railroad tracks. After David's mother died in 1891, his father visited the family in Texas. Jacob was shocked by their poverty and to see his eldest son's only work was scrubbing railroad engines. Unlike so many others in the grip of poverty at the time, David and Ida and their three boys had the safety net of a successful and loving patriarch.

Back in Kansas, Jacob's creamery business had grown into Belle Springs Creamery. He had constructed a new, larger plant in Abilene. To return, David was offered a position as a refrigerator mechanic at fifty dollars per month. He and Ida moved into a small one-story house in Abilene, and their family continued to grow. When Abraham moved to Oklahoma in 1898 to enter the ministry, Jacob bought David and Ida Abraham's larger, Queen Anne–style home on nearly three acres.

By 1898, Abilene had slowed from a bustling cow town to a sleepy prairie town. While a certain western flare still defined it visually, such as its dirt-covered streets and wooden sidewalks, Abilene's wild days were behind it. A bastion of Protestant fundamentalism, Abilene became the cradle of Kansas Prohibition. The community was a prototype of the conservative-leaning Midwest that would become known as the "American heartland."

The Eisenhower family's religious ardor began each day with a Bible reading by David. He and Ida ultimately left the River Brethren for the International Bible Students Association, a precursor to the Jehovah's Witnesses. The Eisenhower boys—numbering six now, one having died in infancy—largely tolerated religion but looked for their way out of it. "Mother," Dwight later recalled, "was by far the greatest personal influence in our lives."[4]

Dwight's youth was spent in successful—if unspectacular—study and hard labor at Belle Springs Creamery. He spent two years first as an ice-puller, shifting three-hundred-pound slabs of ice in the freezer, then time as a fireman stoking the plant's furnaces, before being promoted to night superintendent. "Little Ike," as the family called him then, played baseball and football. Dwight discovered a future in the military largely by chance when, in 1910, he renewed a friendship with Everett "Swede" Hazlett, son of an Abilene physician. Swede had left town to attend prep school in Wisconsin, but returned while he prepared to retake the entrance examination for the US Naval Academy at Annapolis.[5]

Swede told Dwight about the free education available through the Navy and Army academies and how he believed they offered the best ticket out of their sleepy hometown. "It was not difficult to persuade me this was a good move," Dwight later recalled. He took up the project of writing to his senator to gain an appointment to either Annapolis or West Point. His letter ended humbly: "If you find it possible to appoint me to one of these schools, your kindness will certainly be appreciated by me." Already beyond the cutoff age for Annapolis, Dwight faked his application. On January 13, 1911, at the age of twenty, Dwight D. Eisenhower passed the West Point entrance exam.[6]

However, a path that appears simple often turns out to be anything but.

When young Eisenhower surveyed the Kansas prairie and wondered what possible life and future lay beyond it for him, his mother said simply, "It is your choice."[7]

In the ensuing years of his young life, every choice Eisenhower made contributed in some way, large or small, to the creation of the persona that we today refer to as simply "Ike." Coming of age at a unique moment, Ike became a personification of the young nation he would one day lead. Conversely, by tracing a bit of his young life we are able to see the intrinsic ways in which Ike defined the nation.

This is the story of a road trip—a road trip in 1919 of dozens of military vehicles convoying coast to coast on the Lincoln Highway, a road trip so unique and ambitious for its time that the Army dubbed it the First Transcontinental Motor Train. And this is the story of how that road trip marked a perfect convergence: it was both a formative moment in the life of a young man who would become a world leader and a defining moment in the world's relationship with energy.[8]

Gettysburg, Pennsylvania, 1917: Could the FT-17 tank alter
the course of World War I? Trained on the famed Gettysburg
Battlefield by a young Dwight Eisenhower (above, left), these
tankers sought to modernize warfare and to drive the European
front from its entrenched stalemate.[9]

CHAPTER ONE:

Juggernauts of Combat

A lthough the nation—and, in fact, the world—was undergoing a major change in the way people travel, for his first trip east of the Mississippi River in 1910, young Dwight Eisenhower had little choice in how he'd get to West Point, just north of New York City; like most Americans of the era, he took a train.

Not that long before his trip, the technology behind trains had been considered cutting-edge. During the nineteenth century, coal-powered steam rail travel had brought far-off locales such as Kansas and all of the American West into the social and economic orbit of the young United States. The sprawling, disparate nation was made one by the joint government and industrial effort to systematize railroads powered by fossil fuel, and when the transcontinental system was completed in 1869, it was a symbol of national progress.

At the turn of the century, though, the modern world was bringing change at an unprecedented speed. The cultural critic Karl Marx summed up the era: "All fixed, fast-frozen relations, with their train of ancient and venerable prejudices and opinions, are swept away, all new-formed ones become antiquated before they can ossify."[10]

As he rode east to start his military training, Dwight must have been eager to arrive, now that he'd made a choice about what he wanted to do. At each stop along the way, as the train pulled up, he would have seen horses and carts, but also cars, to pick up passengers leaving the train. He wouldn't have known it yet, looking out the window of the train, but he was going to play a major role in shaping how Americans travel by car, right up to this day.

Dwight's plans had begun with college; he had intended to go to the University of Michigan and hoped to play football and basketball. He was now nearly six feet tall and 170 pounds of broad, strong muscle. Even when he followed his friend Swede's advice and instead of college finagled his way into a military academy, Dwight's physical prowess was a big part of his rationale.

Over the years, he developed his speed on the football field. When he had arrived in the fall of 1912, he moved from Army football's line to its backfield. He was a star player that year, until he twisted his knee in a game against Tufts. In practice during his recovery, the knee entirely gave out. He tore the tendon and cartilage and, to his horror, was placed in a full-leg plaster cast. Months later, when the cast was removed, doctors informed Dwight that he would never play again. In the despondency that followed, his studies suffered. But this personal disaster would lead to a turning point in his life. Dwight's former coaches were impressed with his intellect, and, seeking to keep him involved in the sport, they put him in charge of the junior varsity team. Dwight excelled in his new role and began to develop as a leader.

Dwight graduated from West Point as an officer in 1915. As one biographer writes, "He had drifted into the Academy and, as a student, drifted through it." Certainly, however, Dwight showed dedication when he was motivated to do so. During these years, he was fully committed to spending time in the company of young women—a commitment that stayed with him after graduation.[11]

Dwight noticed Mary Geneva Doud—known as Mamie—shortly after his arrival in south Texas, where he'd gone to begin his first military assignment at Fort Sam Houston. She had just arrived for the season from her family's home in Denver, Colorado—her well-to-do family always escaped the city during the hot months. Dwight and Mamie first met by chance, mingling with other young people on the Texas street.

"He's a bruiser," Mamie later recalled, as her initial impression of Dwight. "He's just about the handsomest male I have ever seen." The day after their first meeting, Mamie's maid informed her that "Mr. I-something" had called every fifteen minutes all afternoon. Dwight and Mamie dated over the ensuing months, and on Valentine's Day, 1916, he proposed and she accepted.[12]

In this rapidly changing world, there was no time to delay such plans.

INDEED, THE modern world pressed on every youth in 1916. All aspects of life swirled with radical possibilities, right down to the basic and essential ideas of movement. Consider the magnitude of it: each young American had been born into a world bound by the severe limitations of animal or human-powered movement. Now, rumors circulated of engines that could actually propel large vehicles. This was an era of trial-and-error technologies, where failure was simply a step on the pathway to innovation. The youth were coming of age at a time when tomorrow could differ drastically from today.

In the Abilene that Dwight had known growing up, street traffic consisted of wagons pulled by horses, part of a long tradition of personal transportation. In the late 1800s, industrialization brought mass transit to urban areas. The efficiency of a bus or a subway car made perfect sense in many cities, particularly in Europe, but the vast size of the United States and its sprawling American residential patterns meant personal transport was more practical and popular, particularly in rural areas. But by *horse*? That mode of travel was quickly becoming a symbol of the past and faced an uncertain fate in urban areas.

While modern engineering took some time to trickle out to the Kansas plains, in 1898 the streets of New York City were bizarrely different from just a few years prior. A description of the city from the *New York Sun* must have read like science fiction to people at the time: "At that busy corner, Grand Street and the Bowery, there may be seen cars propelled by five different methods of propulsion—by steam, by cable, by underground trolley, by storage battery, and by horses." All on one street corner![13]

On that busy intersection, the observer makes no mention of the internal combustion engine-powered fleet of vehicles singularly traveling American roads during the next century. The transportation changes in 1898 began without the commodity of petroleum, which remained primarily an industrial lubricant or illuminant. A subsequent shift in powering transportation would eventually change all of that—but it was an alternative shift from the technological free-for-all already in action at the corner of Grand and Bowery in 1898.

At the time, there was every reason *not* to pursue a shift to fossil fuels for personal transportation. Consider for a moment the daunting infrastructure it would require. The massive quantities of oil emerging onto earth's surface were just the start of an unknown and completely undeveloped system: the product had to be developed and systematized. If gasoline power were to be used for personal transportation, endless questions needed to be answered: Was there enough of it? How would it be distributed safely to consumers? Could a device be developed that did not explode in flames when it started?

Tinkerers competed with serious entrepreneurs to craft the method of transport that would move most Americans about their nation. Electricity competed with biofuel engineering to power a wide variety of the new vehicles. But, all of a sudden, when it was discovered in vast quantities near Spindletop, Texas, in 1901, petroleum overwhelmed the energy market.

Although the technology for the internal combustion engine (ICE) was largely dictated through legal restrictions (ironically, the patents were controlled by bicycle and electric vehicle manufacturers who wished to squelch development of the ICE), entrepreneurs doggedly pressed the envelope for radical change—and none more so than Henry Ford. Alone, of course, Ford may not have changed the trajectory of human movement. However, his efforts combined with the gushing crude coming from Texas and elsewhere to jolt a clearer transition away from all other possible fuels.

Ford, probably best classified as a tinkerer, formed his own company, the Henry Ford Company, in 1901. Technology was his focus. Particularly creating fast racing cars. Contrary to most of the transportation marketplace and in defiance of patent controls, he focused on an ICE engine that burned gasoline, made from petroleum. This focus made Ford's path even more challenging. Manufacturers of a variety of vehicles filed persistent litigation that slowed his progress. In fact, Ford's investors repeatedly backed out and forced him to restart his company twice. Of particular frustration to some of them was his refusal to place profit over the price and quality of his vehicles. Ultimately, Ford even had to abandon the company that was named after him—and as a result the Henry Ford Company was renamed the Cadillac Automobile Company. In fact, Cadillac remained a competitor throughout his career.

Despite his difficulties, Ford developed a unique vision that prioritized enabling everyday Americans to own cars—a vision in which Dwight would play a major role, not long after Ford's work took off.

"I will build a car for the great multitude," Ford proclaimed. In his efforts to convince investors, the inventor put his vision more fully in this fashion: "The way to make automobiles is to make one automobile like another automobile, to make them all alike, to make them come through the factory just like one pin is like another pin when it comes to a pin factory, or one match is like another match when it comes to a match factory." This was a radical proclamation—particularly when vehicles operated culturally as a sign of wealth and social standing. Strategically, it was genius. If he made them inexpensive enough, Ford's vehicles might simultaneously take advantage of the high-price of most other vehicles and also of the challenges facing other methods of transportation, including electric batteries, steam power, and mass transit.

Ford trusted in a future of automobility—particularly if he personally profited from each vehicle's manufacture and sale. Ford's sales remained consistent until the 1906 to 1907 season when they quintupled to eighty-five hundred—and that was with the release of the Model T. He was able to reduce the price of these ICE-powered vehicles because of his success with assembly-line mass production—a process that he designed after studying the disassembly lines used in meat processing plants. With the Model T, Ford lowered the price of the automobile below $1,000, allowing personal transportation to become conceivable on an individual basis, particularly in rural and emerging suburban markets. According to historian Tom McCarthy, "Of the first million Model Ts that Ford sold, 64 percent went to the farm and small town market."[14]

In addition to its wide availability, the Model T could also be appreciated for its capabilities: fairly simple repair thanks to

standardization of parts; high clearance, which was a most critical selling feature during the roadless era; and a twenty-horse-
power engine and weight of twelve hundred pounds, which
allowed drivers to get between twenty and thirty miles per gal-
lon of gasoline. With fifteen million sold in the first two decades
since its release, the Model T—nicknamed the "flivver" and the
"tin lizzie"—transformed human transportation. This vehicle
was likely more responsible for the development of large-scale
motoring than any other car in automotive history.[15]

In the early 1900s, in Dwight's hometown of Abilene and
the vast majority of the United States, vehicles like the Model T
were unicorns—turning heads when they passed along dirt roads
shared by wagons, horses, and bicycles.

There was certainly a sense of awe for this new mode of transportation, but it wasn't just a matter of technological advancement. The American consumer was drawn to this movement
because it celebrated one of the founding principles of the nation:
a spirit of autonomy or independence.

At the dawn of the 1900s, Americans had already begun to
envision—and even expect—a future with individualized transportation. And this cultural moment would have a great impact
on Dwight Eisenhower, the young enlistee.

WHEN DWIGHT received his first assignment at Fort Sam Houston
in 1916, the flux in everyday human life was palpable, spurring
both excitement and anxiety, and not just because of changes in
technology. The world had descended to global war in 1914 and
the conflict was ongoing. In his personal life, though, Dwight
by 1916 had found stability and a sense of security through his
marriage to Mamie.

Coming of age in this tumultuous moment, any soldier hun-
gered for the opportunity to participate. Eisenhower was no

different. Once his training was complete, he and Mamie eagerly awaited his assignment to the front. He was convinced that this conflict was his moment to test himself, to see if he'd shine or not. He would soon be faced with disappointment, stuck in the United States, despite his urge to fight. Yet, as a soldier, he was poised to do greater things than he might have ever imagined.[16]

Looking back at the life of this young man from Abilene, it's clear that circumstances can take an ordinary beginning and make something extraordinary. The list of achievements of Dwight Eisenhower, who became known as Ike by his generation and those that followed, is expansive. It is interesting, though, that on that list—which includes leading the Allies to victory in World War II and presiding over the United States for two terms—is rarely included his groundbreaking role as catalyst for the greatest era of energy consumption in human history.

THE RENAULT FT-17 surged up the familiar hillside; it hit the crest, and the strain of the diesel engine waned. Inside, two tankers—soldiers—manned their supremely compact stations: one used the steering lever and pedals to control the caterpillar tracks, and the other stood behind him, manning an extended machine gun, the only weapon on the vehicle. Yet, thanks to its iron cladding and ability to cover any type of terrain, albeit slowly, this French-made Renault FT-17 was a pioneer of modern tank warfare.

On this hillside in autumn 1918, though, the warfare evoked by the terrain was certainly not modern. These were the famed fields of the 1860s American Civil War in southern Pennsylvania. The Battle of Gettysburg, that turning point of the conflict between Union and Confederate forces, took place on these subtle undulations July 1 through July 3, 1863. The Renault

The tank, powered by diesel fuel, revolutionized warfare after World War I and was first integrated into U.S. training at Camp Colt.

tank was creeping by Bliss Farm, where soldiers had engaged in hand-to-hand combat on the second day of that 1863 battle. In the shadows of soldiers who had fought decades before, these tankers were training at the Army's tank school, known as Camp Colt. Young recruits were eager to learn about this cutting-edge, diesel-powered technology that was profoundly shaping the waning days of the European front in the Great War. And there's one more layer of historical significance to this scene: by 1918, the training at Camp Colt was overseen by a young army captain, Dwight D. Eisenhower, in only his second military assignment.[17]

Dwight's initial visit to the historic Gettysburg Battlefield had come a few years prior. In 1915, he had joined fellow "Firsties" (the term still used for West Point cadets in their senior year) for a staff ride to Gettysburg as part of their training. They studied the battlefield and the strategic choices made by commanders in 1863—an Army tradition that began around 1900.[18]

Despite the opportunity to see the battlefields in Gettysburg, Eisenhower was more interested in ancient societies than in the American Civil War. "My first reading love was ancient history," Eisenhower later wrote. "Since those early years, history of all kinds, and certainly political and military, has always intrigued me mightily." And yet the shadows of war at Camp Colt wouldn't have seemed ancient at that time; the Battle of Gettysburg barely qualified as history in the early 1900s. "If, in Abilene, I never became as involved in the Civil War," Eisenhower recalled, "this was because it was relatively recent." Dwight had been born only twenty-five years after the Civil War's conclusion. In fact, the narrative of the war—its storyline—was still being determined. By contrast, the more ancient stories possessed the clarity of purpose that Eisenhower craved. "Closeness to it in time made that war appear commonplace to me," he recalled. But "in any event romance, adventure, and chivalry seemed characteristic of the conflicts of earlier centuries." With remarkable prescience, Eisenhower explained, "Had one of our Civil War veterans [in Abilene], for instance, suggested that not many years later, I would visit Gettysburg to study the tactics of the great battlefield where he had fought, my reaction would have been—'Me?'"[19]

Regardless of the gulf that separated the tactics used in the Civil War and those that Eisenhower would implement in World Wars I and II, there was one important commonality: human leaders still determined the action in battle. From the start, the young soldier was entranced by the gravity of individual choices. He studied the decision-making of previous leaders to prepare himself to play an important role, he hoped, as a soldier— possibly, someday, even to make critical decisions himself.

Eisenhower and other young men yearned to prove themselves on the field of battle, though the environment was nowhere near desirable. On fields of Verdun or Saint-Mihiel, World War I's crashing of old and new created hellacious fighting and living conditions. Trenches were the war's most famous detail, and they grew from a devilish combination of strategy and intentional stalemate. Even if they were complex and reinforced, trenches were essentially an age-old tactic of digging and hiding soldiers in holes so that they can outlast an enemy. Gas, a terrifying new technology, was used in an attempt to flush soldiers out of their hidden holes. It killed humans with a reckless fervor and permanently maimed most survivors, its horrific impact inspiring some of the first rules to rein in the application of modern technology on the twentieth-century battlefield. Even in a time of war, went the logic, certain restrictions were required. Still, mass casualties were one of the defining outcomes of the Great War, with fifteen to twenty-four million deaths and twenty-three million wounded military personnel in the four years of fighting.

Of all the contraptions unveiled on the battlefields of World War I, the tank—among them, the Renault FT-17—was one of the strangest in appearance. Tanks didn't radically alter human capabilities. We weren't suddenly flying through the air or throwing flames, innovations that led to other wartime developments at the time. The tank was an astonishing barricade, so nearly immobile that it needn't qualify as a vehicle. Instead, similar to the great, lumbering elephant warriors used in India's battles of the distant past, the early tank offered the capability of overwhelming a force that remained firmly grounded on earth—and it was used as a tactical response to entrenched enemy soldiers. Its job was to dig into the earth and drive out any soldiers hiding below the surface. At times, the tanks seemed impregnable, able to foil any out-of-date combat weapon, whether spears or musket balls. Yet, also like the great beasts, tanks were unruly and cumbersome. Their use on the battlefield created a raft of complex new concerns.

Trenches were no match for tanks. Once a tank was placed in proximity of trenches, it could securely break through fencing and barbed wire. Tanks had very few ways in which enemies could resist. Instead, their primary enemy was their own technical failures or shortcomings. Also, there was the matter of quantity. Tanks were in short supply, and it was difficult to transport them great distances. Still, when they were employed, they struck fear into the enemy.

The first tank designs are attributed to the British military and were championed by the young Winston Churchill in his role as First Lord of the Admiralty. In February 1915, the British government formed the Landship Committee, a group of military engineers and officers from the army and the Royal Navy Air Service, to review plans for new armored vehicles. Churchill's early efforts to develop the device were run in secret—code-named "tank," from which it derives its name. Known as the "Little Willie," the first tank prototype carried the name of one of its designers. This design led to the Mark I design (you guessed it, known as "Big Willie"). The Willies were first used in 1916 at the Battle of the Somme. Of the four tanks put on the field, the Germans incapacitated two.[20]

Adjusting from this initial experience, British military leaders used a squadron of nearly five hundred tanks to break the German line at the Battle of Amiens in August 1918. Following the war, Britain expanded its efforts to develop tanks and by the time of World War II the device was known as the Churchill tank and carried a six-pound gun. Their integral use in the infamous failed British effort to storm Dieppe in World War II moved the then prime minister to comment, "That is the tank they named after me when they found out it was no damn good!"[21]

In 1917, the French military debuted its first tank, the Char d'Assaut Schneider CA1 Tank, at Berry-au-Bac. The French tanks suffered great losses because of a critical design flaw: forward-located fuel tanks. French engineers came up with an improved design with the light Renault FT-17 tank. It was also the first tank to have a rotating turret.[22]

Early tanks marked a significant leap in battlefield technology, and military engineers would spend decades after World War I perfecting their use from a technical and strategic standpoint.

And, across the Atlantic, the young soldier from Abilene, Kansas, was now at Camp Colt and poised to play a defining role in the mechanization of the US military.

DURING THE winter of 1918, the young army officer Eisenhower watched his post in Texas as others departed for war ("I could see myself, years later, silent at class reunions while others reminisced of battle," he recalled) and grew more and more anxious. "My elation, then, can well be imagined when I received orders in late winter to report to Camp Meade, Maryland, to join the 65th Engineers. This, I was told, was the parent group which was organizing tank corps troops for *overseas* duty." So Eisenhower returned to San Antonio to see Mamie and their newborn son, Icky. Together, the young family awaited what they believed would be his inevitable orders to ship out to the dangerous European front.[23]

At Camp Meade, Maryland, the officers focused first on the 301st Tank Battalion, Heavy, of which Eisenhower recalled, "These men were to man the big tanks, a rarity on World War I battlefields where even the small 'whippets' were not common." Although the tanks had not yet arrived, Eisenhower called them "juggernauts of combat, titanic in bulk even though snails in speed." He was swept up with the same fervor as fellow Americans who saw the tanks as "an irresistible force that would end the war. The men dreamed of overwhelming assault on enemy lines, rolling effortlessly over wire entanglements and trenches, demolishing gun nests with their fire, and terrorizing the foe into quick and abject surrender." This small group of specialized, mid-level leaders would be divided into those who would command the first American tanks on the battlefield and others who would remain in the United States to train the tankers who would staff them.[24]

In Virginia, even though they hadn't yet trained with tanks, the 301st was told that it would soon ship out to the European front with Eisenhower in command. Eisenhower described his euphoria, but also his efforts to control it: "As a regular officer, I had to preserve the sedate demeanor of one for whom the summons to battle is no novelty." Within two days, though, his prospects had changed dramatically. "My chief said he was impressed by my 'organizational ability,'" he wrote later with disdain. "I was directed to take the remnants of the troops who would not be going overseas, and proceed to an old, abandoned campsite in Gettysburg, Pennsylvania, of all places. . . . My mood was black." After his initial visit in 1915, Eisenhower would now be going back, almost three years later—grudgingly—to have the next chapter of his life in southern Pennsylvania.[25]

The Department of War needed to identify strategic locations for military training in 1917, and among them were two sites of Civil War battles: Chickamauga Battlefield in Tennessee and Gettysburg. The headline in the *Gettysburg Times* from May 14, 1917, announced, "Gettysburg Will Get Large Camp for New Troops." At this point, though, the camp had yet to be named.[26]

The local community had already been seeking ways to support the war effort when news emerged that training camps would be established in a few locations in the eastern United States. Despite the large number of local residents in Gettysburg with German heritage, there was no public open affinity for the Central Powers. Local leaders had actively trumpeted the town's strengths, with an eye to military needs—with the hope that a training camp might spur local business. They knew that their federally controlled expanses of land could easily be converted to a training ground; in fact, it had already been used for various military groups. At the local college, the ROTC cadets had even dug trenches to simulate life on the western front.

A few hundred new cadets from Camp Meade arrived at the

new and unnamed training site in Gettysburg on March 16, 1918, and Captain Eisenhower took command on March 29. Eisenhower only stayed briefly before returning to Meade, so that he could prepare the 301st for its deployment. Once the 301st had shipped out to Europe, he, Mamie, and Icky moved to Gettysburg. During their stay, they lived in a series of homes, and one was the Alpha Tau Omega fraternity house at Gettysburg College, which students had vacated in May. Unlike the infantry camp that had preceded it, the new cantonment in Gettysburg received an official name later that year: Camp Colt (honoring Samuel Colt, inventor of the mass-produced six-shot pistol).[27]

Eisenhower threw himself into his immense responsibilities as the nation's leading specialist in the use and battlefield application of the new tank technology. While the experience of establishing a camp for general training and mobilization was fairly familiar to Eisenhower, the type of training that would best serve "tankers" was entirely unknown, particularly since the fighting applications of the tank were being realized on the warfront. And he already faced one big training challenge: no tanks! One tank, the Renault FT-17, would arrive months after his training had already begun, and that was it.

Located near Big Round Top on the battlefield at the center of historic action on the battlefield, Camp Colt occupied the same ground as the infantry camp and was intended for approximately four thousand soldiers. Historian Mark Snell reports that "the Tank Corps attracted some of the best of the army's existing soldiers and new recruits," and the camp soon swelled to sixty-four hundred men. Eisenhower kept the men housed in tents, and they endured all weather, including a challenging, unexpected April blizzard. It was the largest group of men he had overseen, and he was in charge of the entire endeavor.[28]

Even the sacred battlefield gave for the cause. Long before preservationists debated whether or not to cull the forest or mow the grass, the necessity of preparing the tankers for war took precedence over all else. Camp Colt added hundreds of structures, herds of horses and mules, trucks, cars, and motorcycles—and eventually a tank—to the historic area near Devil's Den. To keep down the dust, the military applied eighteen thousand gallons of coal tar (known as Tarvia) to the battlefield's roads. And the sewage from the soldiers corrupted many of the streams on the battlefield. Most, though, viewed these changes as a small sacrifice for the war cause in 1918.[29]

Although the war effort aroused the passions of many young American men, the nature of tank warfare particularly enticed those with acumen in mechanics and engineering. They'd been invited to join in a new era, unlike anything before it. An article in the popular magazine *Motor Age* spoke directly to young and aspiring recruits:

> If you are one of the mental and physical aristocrats of American citizenry—if you are the owner of a sound body, a clear head, an executive, mechanical mind—if you seek vigorous service—if you want to ride to Berlin instead of walk—if you want to do utmost damage to the Hun—join the tank service—it will give you experience, rank, training, and excitement to your heart's content. . . . [The Tank Corps] still requires 3,000 tractor drivers, heavy truck drivers and motor car engine mechanics who have physical strength, mental strength, and executive ability—men without a single physical defect, who can think quickly and clearly and if need be, think for others. These men can be between eighteen and forty-one years of age.[30]

Waiting to usher the recruits into this new motor era was Captain Eisenhower and the infant mechanized approach to warfare.

Eisenhower ran the military camp with an eye to preparing the soldiers for a variety of overseas services. There was a drum, bugle, and fife corps and a camp newspaper called *Treat 'Em Rough*. With authorization from his superiors, Eisenhower also launched an officers' training school, emphasizing rifle marksmanship, drill and ceremony, tactics, and telegraphy. There was a particular focus on machine-gun and direct-fire cannon operation. Eisenhower also oversaw efforts to control the (often connected) military scourges of drunkenness and venereal disease. Building on this matter of military behavior, Eisenhower led Camp Colt's positive relations with the community and business members.[31]

Calm and careful in his intellectual approach, Eisenhower's primary goal was to create a training regime for the tankers—a task that took great energy and creativity, particularly with no tanks available, even imminently. Tanks were scarce, and the few that were available were needed on the warfront. So Camp Colt, the nation's only tank training site, would not have an actual tank for more than a year.

Despite the odds, the Camp prepared multiple tank battalions (the 303rd, the 304th, the 305th, and the 328th through the 338th) and tank centers, which served for administrative and logistical command (the 304th and the 309th through the 314th). Without actual tanks, the soldiers trained on "Battling Lizzie," a moving mock-up based on the Mark V heavy tank and mounted on a truck chassis, which the resourceful Eisenhower had fabricated from sheet metal. Even though they were using a faux tank, the men quickly grew confident with this new technology. Battling Lizzie marked a great testimonial to Eisenhower's ingenuity and leadership.[32]

Unable to replicate entire vehicles, Eisenhower also focused on a key component. He acquired tank cannons and had them attached to trucks. He commented, "The only satisfactory place for firing was Big Round Top." On July 6, 1918, fifty years after the famous battle on Camp Colt's training fields, the *Gettysburg Times* reported, "Some target practice this week with machine guns back of the Round Tops gave the impression that there was a regular battle under way there, but it was only a little demonstration."[33]

At Camp Colt, it was essential for soldiers to study mechanics. They needed to learn about machine maintenance, as the tanks were so prone to breaking down. Recruits for the camp were expected to have some acumen and interest in machine operation and repair. With their rigorous engineering preparation, Camp Colt's trainees, Eisenhower argued, were among the Army's most intelligent and educated.[34]

THERE WEREN'T many tank experts during the World War I era. Colonel George S. Patton was an early adopter of tanks on the battlefield and established an American tank school in Bourg, France. His superior was Colonel Samuel Rockenbach, who served as commander of the AEF (American Expeditionary Forces) Tank Corps. Back in the United States, Colonel Ira C. Welborn was tasked with establishing a tank-training plant that would provide replacement soldiers for the tank battalions of the AEF. To execute his goal, Welborn chose the young Eisenhower. With unparalleled immediacy, this tight brain trust thrust this new technology, with all of its known limitations, into its decisive role in the conflict.

Eisenhower later wrote that he and Patton understood that tanks were intended to "precede and accompany attacking infantry. The prescribed distance between the deployed line of tanks and the leading infantry wave was about fifty yards. The immediate battle task of the tank was to destroy machine-gun nests." He and Patton believed that a high-speed version of the tank would be much more effective, as he later wrote: "We believed that they should be speedy, that they should attack by surprise and in mass. By making good use of the terrain in advance, they could break into the enemy's defensive positions, cause confusion, and by taking the enemy front line in reverse, make possible not only an advance by infantry, but developments of, or actual breakthroughs in, whole defensive positions."[35]

In order to exploit their advantage, early tanks (which often had no guns) needed to be delivered near to their point of action. It was impractical to drive them any long distance. Powering the tank in this early moment of mechanized movement was a four-cylinder, thirty-nine-horsepower gasoline engine that could travel at four to six miles per hour (approximately forty miles per tank of fuel).

Operation of the earliest tanks began externally with a hand crank (though the crank could be operated internally if the tankers were under fire). Communication between the driver and the gunner was essential; however, with no electronic system available for it, soldiers relied on a series of signals by lightly kicking the driver in the back to indicate direction and speed, particularly due to the deafening engine noise on the interior of the tank (compounded by the nearly one-inch-thick armor surrounding the main compartment).

Although no heavy tanks were ever delivered to Camp Colt, a Renault FT-17 light tank arrived at the western Maryland depot in Gettysburg on June 6, 1918. With sarcasm, Eisenhower remarked afterward, "We had not expected to see one until we reached Europe." Particularly for townspeople, though, there was no overstating the otherworldly quality of the high-tech device. "The long expected tank is here," reported the Gettysburg Times. "Camp Colt's officers and men are as happy as a playground full of children with a new toy. . . . It was soon unloaded and driven through town while scores of people watched it with the greatest interest."[36]

This Renault tank was built in France and was first delivered to the Maxwell Motor Company (forerunner of Chrysler) of Dayton, Ohio, for study so that the company could help to fulfill the army's order for forty-four hundred units. Typically, in this era, tanks would arrive to a post without weapons and then be modified in one of two ways, using military lingo of the day: a "female" tank would be armed with a Hotchkiss machine gun; a "male" tank would be armed with a one-pound (thirty-seven millimeter) cannon. Camp Colt's tank was fitted out as a female.[37]

Once the Renault had arrived, drivers used the obstacles around the destroyed, historic Bliss Farm to mimic the terrain of the European front. For expertise on how tanks could be used in battle, Captain Eisenhower brought British advisors Major Philip Hammond and Lieutenant Colonel Frank Summers to Camp Colt; both had commanded tank battalions in the Battle of Cambrai in 1917. "Thus began my connection with allies," Eisenhower recalled. In addition to lending their expertise on this new weapon, the British advisors introduced Eisenhower to the actions and ideas of Churchill, the leader with whom he'd have a critical role in another global conflict years later.[38]

AS EISENHOWER navigated modernization during World War I, he was presented with another challenge: a global pandemic.

Spurred on by the global exchanges necessitated by the war, the Spanish Influenza arrived to the United States in three waves: first, as a mild flu in the late spring and early summer of 1918; second, as a more severe influenza in the autumn of 1918; and, finally, as an epidemic in the spring of 1919. Because the flu moved through the soldiers returning from the war front, Camp Colt, with the first case noted in September 1918, served as an entrepôt for the virus in the countryside of Pennsylvania.

Throughout the autumn of 1918, Camp Colt continued its training activities as the number of infections and deaths from the flu skyrocketed. "In late September," recalled one of the medics at Camp Colt, we "suddenly found our daily chores at our infirmary tents turn into a grim struggle to do what we could to save the lives of our tank corps men. While our buddies overseas were driving relentlessly to victory in the decisive 47-day Meuse-Argonne battle, a sudden onslaught of the deadly pandemic of influenza struck our camp with devastating fury." Eisenhower carried out quarantines of any soldiers running a fever and the administration of new medicines (to which he attributed the fact that his family was spared infection). Tragically, the flu infected hundreds of community members as well.[39]

Eisenhower recalled matter-of-factly, "The whole camp had to be considered as exposed." Quarantining became common, and soldiers could no longer gather in groups. Most importantly, though, Eisenhower made every possible effort to confine the flu to the camp and to keep it out of the town of Gettysburg. The camp reported the deaths of 156 soldiers, and the town added more than one hundred others.[40]

Eisenhower also had to contend with the dreadful impact of the proximity between the town and the camp. He acknowledged this reality with a message of thanks in local newspapers, which read,

> To the citizens of Gettysburg:
>   On behalf of the officers and men of Camp Colt, as well as their relatives and friends, I wish to express to you our sincere appreciation of your timely assistance during the recent regrettable epidemic. It is gratifying to note the spirit of cooperation that has prevailed everywhere, and I am sure your kindness and sympathy will ever remain a bright spot in the memories of those who have suffered bereavement.
>   D. D. Eisenhower Major,
>   Tank Corps Commanding[41]

Globally, the pandemic claimed an unprecedented number of lives: fifty to one hundred million between 1918 and 1919 (the estimate is such a broad span because of record keeping limitations at the time). A significant part of the spread was the fairly new phenomenon of so many humans moving between continents; and, of course, this was exacerbated by world war. Over half a million US soldiers died from the flu. Just as the war had appeared to be reaching its end, the specter of the pandemic further destabilized the world.

In late October 1918, Eisenhower was promoted to lieutenant colonel and received new orders. "My orders for France have come," he told Mamie. He was to lead the contingent from Camp Colt to Fort Dix, New Jersey, for deployment overseas. Once in France he would assume command of an armored regiment. He was ecstatic; however, the army was not. He was summoned to Washington and offered promotion to colonel if he would remain at Gettysburg. He respectfully declined.[42]

In November, rumors of an armistice ran rampant. On November 11, World War I ended with Eisenhower still in Gettysburg (Mamie and their son had moved temporarily to her family's home in Denver). "I suppose we'll spend the rest of our lives explaining why we didn't get into this war," he later recalled. "By God, from now on I am cutting myself a swath that will make up for this," he vowed.[43]

After the Armistice was signed, Camp Colt closed within weeks and Eisenhower oversaw its dissolution before shipping out to Fort Dix (where the sole Renault tank joined him) and ultimately to Fort Benning, Georgia. He had missed the war, he would later feel. "I was older than my classmates," he brooded, "was still bothered on occasion by a bad knee, and saw myself in the years ahead putting on weight in a meaningless chairbound assignment, stuffing papers and filling out forms. If not depressed, I was mad, disappointed, and resented the fact that the war had passed me by."[44]

While Eisenhower was considering leaving the service, as many of his colleagues had done, the army settled on Camp Meade, Maryland, as the permanent home for the tank corps.

Although he wasn't necessarily aware of it during those wartime years, Eisenhower had been deeply involved in a massive transition in human life. All around the experience of World War I, new sources of power in the form of petroleum had permeated basic strategy and planning. There was no going back—only forward. As Eisenhower set up the new camp at Fort Benning, the War Department asked for volunteers for a cross-country publicity tour. Eisenhower was among the first to volunteer. "I wanted to go along partly for a lark and partly to learn," he later recalled.[45]

It was his choice.

Washington, DC, the Ellipse in Potomac Park: Although motor vehicles had become more and more frequent around the nation's capital during World War I, the opening ceremonies of the Transcontinental Motor Train gathered an impressive number of military vehicles under the direction of government and industry leaders on July 7, 1919.

CHAPTER TWO:

A Mobile Display Window

"This is the beginning of a new era," Secretary of War Newton Baker announced to the crowd of a few hundred, who stood with rapt attention in the heart of Washington, DC, the morning of July 7, 1919. Absent from the audience was the young Dwight Eisenhower. Although he had signed up to accompany this expedition, he remained in the nearby Camp Meade and would join the entourage a few days later.[46]

In truth, Eisenhower was very nearly not part of this entire excursion. After the war, the military was in a place of transition, which meant there were new career decisions to be made. Mamie and their baby son, Doud Dwight, remained in Denver with her family because Camp Meade offered no married quarters. Although his wartime responsibilities had brought him the rank of lieutenant colonel, Eisenhower was soon to be demoted back to captain, and his pay would drop by nearly a third—to $200 per month.[47]

And, in the interim, one of Eisenhower's subordinates had offered him a civilian job at his family's manufacturing firm in back in Muncie. Eisenhower seriously considered taking the job, particularly because he felt the new Tank Corps might be entirely disbanded in peacetime. When he received word of the convoy, though, he made his choice: to stay on and to join the party. For him, the convoy was intriguing and fun; he appeared to have little awareness that there might be a larger significance to the undertaking.

On that early July morning, Baker christened the event and the era a new one for the human species and particularly for how Americans would soon move about. Around him were futuristic machines and vehicles that had previously only been seen together on the battlefields of World War I.

Camp Meigs had been hastily formed during World War I in northeast Washington, DC, as a base for the US Army's new Motor Transport Corps. Trucks from the camp were frequently seen carrying troops and supplies to various locations around the nation's capital. These trucks weren't all that different from horse wagons; they simply had no horse. The driver would sit on a bench and hold on to the steering wheel for dear life, delivering groceries, ice, or building materials. Often, the trucks carried sculptures that were installed throughout the nation's capital. Few people would have looked twice at a passing military vehicle from Camp Meigs, but the movement on the streets of DC on July 7 that year was unlike anything anyone had ever seen outside of the battle zones of the western front. It was the largest gathering of military mechanized vehicles ever assembled, and it was a vision of the technological future close at hand.

But imagine the startled capital residents as a two-mile-long procession moved from Camp Meigs toward the White House to take formation around the Ellipse, the grand central parklike landscape designed to be the center of the city's system of streets! The procession had followed a carefully planned schedule that allowed for ninety minutes to cross the city. The crude, massive engines were powered by cylinders pounding a deep rumble and chains that carried the torque to four wheels made of stiff rubber, which gave little comfort to passengers. Mostly, these were utilitarian, baseline, no-frills military vehicles. The convoy was composed of Companies E and F of the 433rd Motor Supply Train, Company E of the Fifth Engineers, and Service Park Unit 595. The convoy included a medical detachment, a field artillery detachment, and a small group of commissioned officers as observers of the specialized equipment. Essentially, the observers' job was to operate the vehicles as a demonstration at each new site. Thus, the observers' status wasn't about watching, but was instead a reflection of their lay military activity as the vehicles spent the bulk of their time in tow.[48]

The vehicles assembled with Baker around the monument that would soon be christened "Zero Milestone," and the First Transcontinental Motor Train was transformed from its loud, rattling reality into a mythic symbol of national aspirations. The temporary monument (which was donated by S. M. Johnson, president of the Good Roads Movement, and later replaced with a permanent marker in 1923) was supposed to be the equivalent of ancient Rome's "golden milestone" located in the Forum. Renowned for its first organized use of roads, Rome provided perfect symbolism for this road trip. Zero Milestone would mark the distances of all national highways from Washington, DC. Clearly, from the outset, the organizers wanted this convoy to be the catalyst for a national transformation—even if the young soldiers such as Eisenhower scarcely appreciated it at the time.[49]

A cast of high-ranking officers and other notables gathered around a speaker's podium, with the monument covered in a white sheet behind them. Hundreds of onlookers listened to speeches from many speakers, including President Warren Harding; however, it was Baker who presided, noting,

The world war was a war of motor transport. It was a war of movement, especially in the later stages, when the practically stationary position of the armies was changed to meet the new conditions. There seemed to be a never-ending stream of transports moving along the white roads of France.

One of the remarkable and entirely new developments of the war was the inauguration of a regular timetable and schedule for these trucks. In the daytime they were held in thickly wooded sections, but at night each one started out with a map and regular schedule which was as closely followed as the modern railroad. In no previous war had motor transportation developed to such an extent.[50]

Finally, in closing, Baker made the reference to a "new era" that opened this chapter. After his speech, Baker unveiled the temporary monument, a granite obelisk. He then directed the leader of the convoy, Lieutenant Colonel Charles McClure, to "proceed by way of the Lincoln Highway to San Francisco without delay. . . ." In truth, though, there was little continuity to the "highway" that they would follow. It was a concept at that point, and the organizers hoped that the convoy would play a large role in convincing the public to embrace a transcontinental highway. They had drawn an aspirational line across the map of the nation for the vehicles to follow; however, on the ground it would often be little more than a wagon path. Given the lack of roads and the untested technology of vehicles, particularly at this scope, the organizers may as well have been planning a trip to the moon.[51]

At its starting point in the Ellipse, the organizers placed the temporary Zero Milestone sculpture that would be updated and made permanent in 1920. Adding to the ceremonial pomp of the moment, before the convoy left DC, Representative Julius Kahn of California had presented two wreaths to Colonel McClure that would be transported across the nation and given to California Governor William D. Stephens when they completed the trek.

Without Eisenhower on this first day, the convoy rolled out to the northwest in a grand incursion to the unknown.

The convoy first moved to Gettysburg, Pennsylvania, where they turned left on the Lincoln Highway (LH), the most famous highway of its day. Stretching out of southern Pennsylvania, this roadway had been latched onto by many communities in the eastern United States for the economic prospects that it promised after it was officially established from 1912 to 1913. In Pennsylvania, the LH utilized established roadways that had served as trails in some of the nation's earliest transportation networks. In other regions, it drew new construction and settlement, much of which had been on hold during the war and in the years immediately prior while developers searched for sources of funding.

Washington's *Evening Star* newspaper called the convoy the "longest and most thoroughly equipped and manned Army motor train ever assembled." And it would travel on a highway that, just a few years earlier, newspapers had described as "an imaginary line, like the equator!" Left largely unsaid was that in many areas the road remained imaginary, and it was likely some of the vehicles would be lost along the way. [52]

The *Evening Star* summed up the start of the journey: "This long Army motor train, composed of over sixty trucks and with a personnel of more than 200 men, rumbled slowly out of the city on a journey across the continent shortly after 11 o'clock this morning. It is to be self-maintained and self-operated and carries road and bridge building equipment, so that in case of a wash-out repairs can be speedily made." [53]

Despite the absence of two of its members—Eisenhower and Major Sereno Brett—each of the eighteen observers offered important expertise and experience: often, specific knowledge about a type of vehicle as well as military leadership skills. Describing his own involvement, Eisenhower recalled later,

As it passed through American communities, the vehicles of the convoy symbolized a great leap forward in transportation technology.

I heard about a truck convoy that was to cross the country from coast to coast and [was] immediately excited. To those who have known only concrete and macadam highways of gentle grades and engineered curves, such a trip might seem humdrum. In those days, we were not sure it could be accomplished at all. Nothing of the sort had ever been attempted. . . . I wanted to go along partly for a lark and partly to learn.[54]

Eisenhower liked the fact that the use of all types of Army vehicles would allow for "comparative tests" and that Americans would have the opportunity to see the vehicles that had been used in the war effort, including a small Renault tank that would be carried along on a flatbed truck.[55]

As a typical young lad who was somewhat chagrined and bored with the pomp that the organizers associated with the convoy, Eisenhower explained later with his trademark ironic sneer,

"My luck was running; we missed the [opening] ceremony." To his youthful eye, there was no certainty about what the convoy would mean. But little did he know that the choices that he would eventually make would be primary catalysts in defining a revolutionary era in human living and, in hindsight, to solidify the importance of the plodding convoy.[56]

With the publicity mill churned by the convoy's organizers, there was swell of attention on the caravan of vehicles moving its way out from its point of origination in DC, particularly in military towns. "Did Not Stop Here" read the disappointed headline in the *Gettysburg Times*, the town that certainly had earned its military stripes first in the 1860s and then again with Camp Colt during World War I. The bypass of Gettysburg was simply a matter of timing, but the *Times*'s reaction shows that American communities yearned to host the convoy and to participate in this moment—whatever it would become in the annals of history.[57]

For Eisenhower, who would largely act as an observer and technical guide during the trip, the appeal of the First Transcontinental Motor Train was mostly about a lack of better alternatives; understanding the motives of its organizers and for the nation, though, requires a bit of unpacking.

Indeed, the true importance of the convoy only emerges when we see the energy abundance that followed. To reach a remarkable era of consumption, the nation needed to transition from a limited age of animal and human transportation and enter the unknown frontier of powered-machine travel—replete with all the limitations and difficulties that such a shift required: a century-long trek that Eisenhower himself would largely guide. And the convoy was an early step in that direction. It was an elaborate, complex, and highly choreographed project, designed by a few interested parties, who were determined to prod an American cultural change that would facilitate one of the most significant energy transitions in the human story.

*   *   *

IN THE emergence of early motoring, Packards were significant. They were a state-of-the-art symbol of beautiful design, right down to their chrome detailing. And they marked the centerpiece of the 1919 convoy.

In the first decades of the twentieth century, auto manufacturing was a highly individualized and independent enterprise. Often, an entrepreneur started a brand based on personal priorities and promoted the automobiles with individual flair at events from carnivals to road races, to gain the consumer's attention. These entrepreneurs saw the energy transition from horse and carriage to automobiles as a business opportunity. The convoy was rooted in the same spirit of individualism—just on a very different scale. And the organizational efforts operated with more cohesion than any other automobile promotion that had preceded it. Business acumen, political influence, and determined boosterism fueled the formation of the First Transcontinental Motor Train, and the Packard and its inventor led the way—figuratively and literally.

Why was a drive across America imperative in 1919? It was not because the government or its military decided so. There was no "Inflation Recovery Act," as we've experienced in recent years, through which the government might prod Americans to transition to new modes of life. Certainly, the US military and government supported the convoy, but they were following the lead of thinkers such as James Ward Packard, who had been negotiating the transition to automobiles for years. For Packard, the cross-country road trip was a way to gain the attention of Americans and potentially tip the scale of how they would move about in the future.

Packard began building automobiles in 1899 in an effort to diversify his small electronics business in Warren, Ohio. In 1902, he sold one of his early vehicles to a businessman named Henry Joy. And, when Joy asked, Packard tossed the entire vehicle portion of his business into the sale. Joy brought together the passion of the era's auto inventors, like Packard, with the industrial systems ideas that were forming in Detroit, Michigan, particularly inspired by the management philosophies of Frederick Winslow Taylor and others. But when Joy's auto plant opened in 1903, its emphasis was less on the manufacturing system and more on crafting each Packard as a complete, effective technological marvel—practically a work of art.[58]

Joy proved to be a remarkable publicist and spent long hours promoting the new Packard, including a 1906 drive from Detroit to New York, which he completed in three days. In 1908, Joy appeared before Congress's Committee on Ways and Means as a representative of the nascent American automobile industry. In this jumbled era of startups, Joy foreshadowed the twenty-first century's Elon Musk. Before the rest of the auto industry had really even taken form, Joy advertised his Packards with flair, displaying them in flamboyant showrooms amid elaborate celebratory events. By 1910 he had Packard dealerships in thirty-eight American cities as well as Paris. By 1912, his Detroit auto plant employed seven thousand workers, and he had $1.4 million in orders.[59]

"The Packard," wrote the well-known editor Elbert Hubbard, "is a car for a patrician. It belongs to the nobility, and is of the royal line. . . . To own a Packard is the mark of being one of fortune's favorites. It satisfies ambition, soothes aspiration, and gives a peace which religion cannot lend." The company's slogan belied the spirit of exclusivity: "Ask the Man Who Owns One." Even at this early point in the auto consumer marketplace, Joy had found a way to corner the market of educated, affluent purchasers, similar to Tesla's success in the early 2020s.[60]

For a broad cultural transformation, though, the change could not be relegated to the elite; other auto pioneers had to take aim at a product for the masses. In particular, Henry Ford needed to create an automobile that every American might see as affordable.

The auto industry was already showing clear growth. American vehicle registrations rose from 800 in 1898 to 8,000 in 1900 and to 902,000 in 1912. Riding this new wave in the United States, Ford honed his product and filled his coffers to sustain a legal fight over Selden's patent, which restricted the use of the internal combustion engine. In 1911, after extensive litigation, Ford won and was free to produce automobiles without penalty.[61]

Unintentionally, Ford's competitors in automobile manufacturing contributed to his growing business reputation. As a result of the 1911 settlement, many manufacturers entered into cross-licensing agreements that would be overseen by the Automobile Manufacturers Association. Despite this growing inertia behind ICE, though, in the 1910s, America's vehicular future was not yet certain.[62]

Industrial leaders such as Packard and Ford—not consumers—deserve most of the credit for redefining transportation. These leaders, in particular Ford and John D. Rockefeller, almost single-handedly wrestled the energy transition away from electricity and toward a different kind of energy. Ultimately, that is where marketing—and events such as the 1919 convoy—would prove crucial.[63]

During the 1910s, many people believed that commercial vehicles, particularly trucks, were best powered by electricity. As a replacement for the horse-drawn wagon, a commercial vehicle only had to be reliable for short trips, which played to the limits of electrics. In addition, unlike a pleasure vehicle, the delivery truck had to be run in all weather, which again favored electrics since they were a much simpler technology. Also, there was great variability in the speed and acuity with which drivers guided gas-powered trucks. Speeding was considered a major problem at the time—and, in fact, many owners opted to install governors (devices that would cap the vehicle's speed) on their gasoline-powered truck engines.

Efforts on behalf of electric transport continued with vigor as late as 1914. Thomas Alva Edison, the United States's celebrity inventor, dedicated a two-hundred-thousand-square-foot, four-story factory in his West Orange, New Jersey, complex to perfect a battery that could last more than forty thousand miles in cars and serve for various other duties as well. The existing manufacturers of batteries, though, did not wish to see Edison's Type A battery succeed. In addition, although the battery had been successful in trucks and other uses, Edison wasn't interested in manufacturing vehicles. Simultaneously, writes Edwin Black, Henry Ford in his own notes came to a quiet realization about his own innovation: his vehicles possessed "a faulty electrical ignition system to drive the pistons." With the settlement of the Selden patent suit, Ford no longer needed to steer entirely clear of the concept of electrically powered transportation. So, in the fall of 1912 Ford joined forces with Edison to further revolutionize transportation.[64]

Edison's broader experiments brought storage batteries into American homes, beginning with his mansion in Llewellyn Park, New Jersey. The September 1912 *New York Times* announcement stated that Edison had perfected "a combination of gasoline engine, generator, and storage batteries by which, for a modest expense, every man can make his own electricity in his own cellar." Among the electric items to be charged in this new home universe would be Ford's Type A–powered electric vehicle. Black reports that Ford had promised to produce 12,500 per month in the first year of production. He writes:

Finally. It was happening. The automobile revolution, which began as an electrical phenomenon, would return to the concept advanced nearly a generation earlier. The world could become a cleaner, quieter, more efficient place, drawing its strength from nature, from electricity. The American spirit of independence would be achieved not only by permitting mobility but by enabling stunning individual self-sufficiency.

On January 9, 1914, Ford released his plan to the public. In it, he promised that each EV would likely be priced between $500 and $700. As part of this EV project launch, Ford also told the public that it would be led by his son, Edsel.[65]

The internal experiments, unfortunately, showed problems with the battery almost immediately. Edison, however, maintained his confidence in public. He told the *Wall Street Journal:* "I believe that ultimately the electric motor will be used for trucking in all large cities, and that the electric automobile will be the family carriage of the future. . . . All trucking must come to electricity." In his laboratories, Edison's batteries tested positively during this time and Dodge also went public with its own effort to release an inexpensive electric car. The competition seemed to be tipping towards the EV when, instead, October 1914 brought the foreboding news of a war on a massive scale in Europe that would alter everything—including the future of human transportation. Each of these developments factored into the death of the American initiative for electric transport, and led up to a fateful event: in the crisp evening of December 9, 1914, with a flash, Edison's complex went up in flames.[66]

Edison's complex was primarily built of fireproof buildings; however, the fire escalated quickly and burned until the following day. Ultimately, only Edison's private laboratory and the storage battery factory were saved. Edison's career would never fully recover, and he suffered a nervous breakdown as a result of the tragedy. As Edison dealt with these personal struggles and World War I loomed, the future of EVs was swiftly extinguished in favor of transportation best suited to the war.

Among the most successful businessmen of the era, Packard, Ford, and other automobile manufacturers supported the war cause while also surveying the business opportunities that would emerge afterward. The convoy was a joint effort between these auto entrepreneurs and the military and government leaders that had catapulted ICE vehicles into use on the war front.[67]

LIKE A great actor in a mundane high school production, the vehicle revolution immediately altered all living patterns that surrounded it. The technology for an actual vehicle marked only one aspect of the massive transition that had to occur. Regardless of what power source was in use, the changes in human movement created new needs and relationships between humans and their living environment.[68]

In the United States, this revolution led to a decentralization of the population; however, this was a gradual shift in where humans chose to live. The first few decades of the twentieth century were instead defined by grassroots efforts to create businesses that would accommodate the needs of the automobile and a new society of drivers—similar to how tech startups integrated the internet and cell phones into everyday living patterns in the early twenty-first century. For such social and cultural transformation, the organizers of the 1919 convoy believed a nationwide demonstration was essential. First and foremost, though, regardless of the economic class of the vehicle owner, each of the pre–World War I vehicles suffered from the same malady: Where could they be driven? Why should consumers flock to vehicle showrooms if a purchase meant being encumbered with an albatross, out of place and difficult to use?

With the first useable, generally affordable vehicles in place, adventurous members of the wealthy class hired a personal mechanic and set out on expeditions throughout the first decade of the 1900s. Regardless of the lack of roads, the holy grail of goals for such excursions was to follow the path of Lewis and Clark and other American pioneers who had transected the continent. To do so, most often, these high-society drivers chose an expensive, reliable work of mechanical art: a Packard.

In 1903, launching a publicity tour for the automaker's new model "Old Pacific," a Packard accompanied by its own photographer (and mechanic) left San Francisco, California, on June 20 and made it to New York on August 21. The successful sixty-two-day trip proved, according to the company, "The land of the Packard is everywhere." This publicity stunt became an annual tradition for years to come.[69]

Henry Ostermann, who was born in Indiana and had worked as a flagman, brakeman, and conductor on the Illinois Central Railroad, became passionate about auto travel and completed his first successful cross-country journey in 1908. His primary skill set seems to have been as an organizer and publicist, and he spoke widely about his own trek. He told an audience at Old Deutsches Haus in 1912 that an amateur driver could expect the journey to take between sixty and ninety days. In the audience, Carl Fisher, an avid cyclist and racer with an acumen for publicity, vowed to try to change the situation. Fisher had made a name for himself in 1904 when in hopes of gaining attention to sell cars, he flew across Indianapolis in a car suspended from a large balloon. After hearing Ostermann, Fisher spoke with friends and associates about his dream for a "Coast-to-Coast Rock Highway" and spurred them to join him in a national crusade, which they referred to as the "Good Roads Movement" (this organization would ultimately sponsor the zero-milestone monument in Washington, DC, where the convoy later began in 1919).[70]

Amid the growing energy and interest around all vehicles—but particularly Packards—the war emerging in Europe in the 1910s also began to transform American industry. Early in 1911, Packard began converting its plant to produce a motor truck for military use. Although it took years to develop the specifications, these discussions initiated a relationship between manufacturers and the military. Packard received its first order in 1914 for 180 trucks to be used by the Allied armies in western Europe. By then, the plant was running around the clock and producing a new truck every forty minutes. With this new emphasis, few Americans spoke openly about cross-country travel—at least for the moment.[71]

In the early 1910s, the infrastructure of automobility was complex, even daunting. For some critics, the required changes were significant enough that Americans might simply decline a transition—instead look the other way and keep feeding oats to the family steed. However, to a growing number of others, such as Packard, Joy, and Ford, the amount of required change marked an economic opportunity and a clear model for social and cultural progress.

The switch to petroleum for personal transportation was the result of aggressive efforts to guide consumer tendencies from the top down. With supplies of petroleum overwhelming the existing need and additional reserves being found at a rapid clip, petroleum's role in American life was ripe for expansion. At this moment, industrial leaders—not American consumers—deserve most of the credit for reenvisioning American transportation.[72]

Some industry leaders were keenly aware that drivers required sustenance. What would they eat? Where would they get a drink? Early roads could not simply be a desert of human services. Tea rooms and coffee houses had become popular in urban areas and were soon a staple of roadside culture during the 1910s. From the Victorian era tradition of high teas, tea rooms had developed—particularly among women—as one facet of high society's emphasis of gender independence. These businesses had become locations to relax and to enjoy a light lunch or afternoon tea. And by 1910, tea rooms were seen by many hotels as more profitable than their bar rooms. It was only natural that in the new economy of roadside development, standalone tea rooms also began to develop. In the early decades of driving, use of cars was a leisure activity, which also synched nicely with the clientele and culture of the tea rooms. Indeed, soon roadside tea rooms had sprung up in rural areas as well.[73]

In 1917, *Good Housekeeping* reported, "Until the automobile was graduated from the class of luxuries into that of necessities, tea-shops were successful only in the larger cities. Today they flourish in the smallest hamlets and flaunt their copper kettles and blue teapots on every broad highway." Historian Jan Whitaker writes, "Often, these were mom-and-pop operations, with mom running the tea room while pop pumped gas."[74]

Unlike tea rooms, the gas stations that "pop" ran did not exist prior to the automobile age. The product of Kerosene was used primarily for lighting and, therefore, was normally distributed through general stores or grocery stores, either in cans or refilled by clerks in storerooms. Once vehicles with internal combustion engines began venturing about, though, they created a new market for a different kind of fuel—and it was essential to offer a supply wherever cars drove.

Gas stations typically sold gasoline, which required a new refining process that grew out of "thermal cracking," a practice that began in 1912 and doubled the amount of gasoline yielded from crude. This product was normally distributed through tanker trucks and train cars to filling stations at convenient points for consumers' transportation. Unlike tea rooms and other roadside points that had been devised prior to the automobile, the filling station was first and foremost a technological innovation—eventually, it became integrated into American culture.[75]

One of the gravest challenges of a transition to the internal combustion engine, though, was how to safely store and transport the highly flammable, explosive gasoline. At this time, after being refined gasoline was normally stored in bulk stations that held the liquid fuel in large cylindrical tanks that rested on wooden frames above ground. From this dispersal point, the gravity-fed system could be moved to horse-drawn tank wagons that might then take it to other dispersal points, including the local livery, auto repair shops, or even dry good stores. In such stores, customers bought gasoline by the bucketful to take to their homes for various uses. This was also the era when another layer of entrepreneurs crafted horse carts that could hold a small tank of gasoline so that the product could be sold directly to consumers door to door. In many homes, consumers kept a measuring can on hand to could keep their ration available. If they were transferring it to a vehicle, they most often employed a chamois-lined funnel to make the final transfer into the auto's tank. As more and more drivers chose to bypass these gas brokers and went directly to the bulk station, demand grew for a new type of business. This flow of traffic also created a great opportunity to organize and systematize roadways, which often became clogged with auto drivers waiting for or searching for a fill-up.[76]

Even in the early 1900s, such an energy transition created a steady stream of new business opportunities. And ingenious entrepreneurs had advanced gasoline dispersal technology significantly by 1905. Most notably, a St. Louis entrepreneur, C. H. Laessig, improvised by standing a hot-water heater on end so that it could be filled with gasoline that could then be directly dispersed through a garden hose and into any vehicle nearby. Soon, this system completely outmoded buckets and the innovative gravity-fed system became the new standard. It did not, however, radically increase the amount of gasoline that was immediately available to consumers. Another innovator, Sylvanus F. Bowser, is credited with devising a pump that could pull the gasoline out of a barrel for dispersal into vehicles. His pump could be utilized by existing outlets, such as grocery stores, but the waterproof cabinets that held the pumps could also be locked and kept secure overnight. The combined force of such innovations powered the energy transition of this era. Soon, the flexible pump technology joined with other efforts to create the template for a more standardized, free-standing "filling station" that started opening that same year.[77]

Truly, Bowser's pump proved to be the vital link in this new energy system and it allowed gasoline consumption to be freed from the gritty and dangerous bulk station. While gas retail outlets and filling stations were a critical necessary step in making gas-powered autos more convenient and feasible for everyday use, they were not without their own problems. Most often the haphazard and unplanned streetside refueling stations led to serious congestion and traffic. The early years of mixed-use roads also frightened the horses that still frequented passed and let to small-scale steam explosions. Planning efforts prioritized community safety and the first guidelines—suggestions would be a better term—came from the Automobile Gasoline Company in the 1910s. Historian Daniel Vieyra describes the approach to land use in this fashion: "On an ordinary city lot, the company constructed a small brick building, paved the yard behind, and erected four gas pumps. These pumps . . . drew not from small, above-ground barrels but from safer, more advanced underground tanks; they combined this convenience with Laessig's hose hook-ups, which funneled gasoline directly into the car."[78]

Placement of the new devices—and their flammable product—were of primary interest to developers. As hard as it is for contemporary drivers to imagine, roads of the early twentieth century were not yet a fixed entity—they varied daily both in quality and in their very existence. One day, a road could be busy and the next it was abandoned and a paved alternative had opened nearby. While such variability made prefabricated stations desirable (stations could readily be dropped wherever they were needed—even in remote locations), their locations helped to create an auto infrastructure from a landscape with little structure. Stations would make automobiling much simpler and provide the vital link tying human transportation into the gravitational orbit of petroleum.[79]

As more and more drivers hit the roads, the demand for roadside food grew too. By the 1890s, we see clear distinctions within this transportation transition: the car needed to first be connected to petroleum supplies to make the mode of transportation viable—this was essential to the technology. Linking human food supplies to automotive transportation only increased convenience; it was not essential to broadening the use of automobiles. Providing food for drivers was just one example of the auto culture that quickly emerged to take advantage of or exploit a new consumer market of drivers.[80]

From an economic community of urban consumers—pedestrians or riders of mass transportation—the public was fracturing by the 1910s, as more and more Americans became independent-traveling drivers. Similar to scattered motes in the universe, automobile-related developments—roads, tea rooms, filling stations, and diners—gathered cultural and economic inertia. The United States that emerged after World War I was ready for growth and development, and the automobile—powered by gasoline—represented one of the most obvious opportunities of the era.

AS LITTLE to no attention had been paid to road development in the United States, the construction and organization of the LH in 1913 marked a national symbol for a brave new era. Here, in our story of the 1919 convoy, we come to what is arguably the primary catalyst for its organization—and, likely, completely beyond the notice of young Eisenhower and the other military observers.

Few drivers today know the LH as more than a name on a map—a name that has been superseded by one route number or another. If anything, it is known in the 2020s as a relic of a bygone era. That is the era, though, to which we must return our gaze for the story of Ike's convoy, the era when the LH meant the cutting edge of transportation. In fact, though, there was no other route that the convoy could follow, moving west through Pennsylvania.

In his Detroit office on April 14, 1913, Packard's Henry Joy said to a group of interested industrialists that their project for a road across America should be known as the Lincoln Memorial Highway and should pick up where the current LH left off. The Lincoln Highway Association (LHA) was formed in the following month with Joy as its first president. The founding directors were luminaries: from the auto business, Carl Fisher, Roy Chapin, and John Willys; from the tire business, Frank Seiberling of Goodyear; and from the cement industry, Albert Gowen. All the raw materials needed for a good roadway![81]

With their passion for developing infrastructure, the LHA looked west to configure a path to the Pacific that was both pragmatic and helpful to development and expansion. Plotting the route and raising funds occupied almost a decade, as they waited for World War I to end; during those years, they often discussed a grand demonstration of the auto age that might christen their road. If publicized effectively, they believed that a road trip would be the single best form of instruction for Americans on this new lifestyle, and the LH would become the primary symbol of the age.

Even though this event eventually became known as "Ike's Road Trip," Eisenhower and the rest of America's youth knew nothing of the planning and plotting of these automobile schemers.

*   *   *

FROM WASHINGTON in 1919, the pilot car of the convoy led the
way. It was the American answer to vehicular luxury: a white
Packard Twin Six. Ostermann drove the Packard, at the head of
the entourage, largely due to the fact that he knew the way—he
had completed nineteen cross-country treks by this point.

The great convoy passed through Frederick, Maryland, bypassed Gettysburg (at least partly because of low-lying covered
bridges), and arrived in Bedford, Pennsylvania. Compared to
many states, the commonwealth was well known for road construction. A landscape organized with the priority that roads
for any sort of travel should extend ten miles out from every
county seat, Pennsylvania was well situated to absorb motor vehicles. Their rapid adoption drove the state's desire for smooth,
hard surfaced roadways throughout the state. Although many

Particularly in the open field, the convoy traveled on roads originally for cattle and horse-powered wagons.

entrepreneurs had planted tea rooms, diners and gas pumps along Pennsylvania roadways, as a military enterprise the convoy carried its own services.

Nearly all of Bedford, Pennsylvania's approximately two thousand residents came out to cheer on the convoy. Dr. S. M. Johnson, the convoy's official speaker, represented the National Highways Association, and said to the crowd, "We are crossing the continent to impress upon all leaders of public action in the world that the next step in the progress of civilization is to provide road beds upon which rapid transit motor vehicles may be operated with economy and efficiency. This is true, not only of backward peoples, but also of the most advanced nations, including our own." To little notice, Eisenhower and Brett had joined the convoy in Frederick, completing the team of officers who would staff the convoy. Of the four speeches in Bedford, Eisenhower later ironically recalled that it was "only a slight taste of the hot air ahead."[82]

It is worth noting that as the convoy set out westward, its official goals were much more specific than Johnson's "hot air." The diarist of the convoy, First Lieutenant E. R. Jackson, began his report with the four purposes of the undertaking:

a) The War Department's contribution to the Good Roads movement for the purpose of encouraging the construction of through-route and transcontinental highways as a military and economic asset.

b) The procurement of recruits for the enlisted personnel of the Motor Transport Corps . . . as candidates for the Mechanical Training Schools. . . .

c) An exhibition to the general public, either through actual contact or resulting channels of publicity of the development of the motor vehicle for military purposes, which is conceded to be one of the principal factors contributing to the winning of the World War.

d) An extensive study and observation of terrain and
standard army motor vehicles by certain branches of
the army, particularly the Field Artillery.[83]

However, entrepreneurs from the automobile industry and
related businesses had imbued the convoy with non-military-
oriented purposes as well.

Indeed, in a particularly remarkable development, the US
Signal Corps supplied a film crew. With a group of journalists,
they traveled in three brand-new Willys-Overland Mysteries (a
four-wheel-drive off-road vehicle) throughout the excursion.
The footage they created served two purposes: shown in cinemas throughout the United States, these very early motion pic-
tures presented the convoy both as a seminal event in American
history and as a demonstration of new vehicle technology—
essentially an advertisement for auto producers.[84]

Let us stop to take complete stock of the historic truck train
as it rumbled out of Pennsylvania: eighty-one vehicles, thirty-
seven officers, and 258 enlisted men. The cast of vehicles broke
down as follows: forty-six trucks, ranging in size from three-
quarter-ton Dodge light-delivery vans to massive Macks with
over five tons of carrying capacity; sedans and other vehicles,
composed of Packards, Rikers, Whites, Garfords, and four-
wheel drives. The officers rode in passenger cars—Whites,
Cadillacs, and Dodges—accompanied by nine Indian and
Harley-Davidson motorcycles that acted as scouts. Two of the
cargo trucks were mobile machine shops, including spare parts
and even a blacksmith shop. With only filling stations only
erratically available, the convoy brought its own: two tankers
that carried 750 gallons of gasoline each. Another tanker
carried the same amount of water. In terms of human support, five General Motors (GM) ambulances were joined by
two ambulance trailers and four kitchen trailers, as well as

Throughout the long passage, the Militor was the device that could extract each of the other vehicles as
they faced innumerable challenges. It was the hero
of the trip!

Eisenhower described the convoy assignment as a "lark," and
participants believed that they were part of a great leap forward
for the nation.

two Trailmobiles (four wheelers) and two two-wheelers called Liberties.[85]

On top of this core of somewhat familiar vehicles, the military included some of its more specialized—even secret—equipment: the Loder, a pontoon trailer that could carry vehicles across rivers such as the Missouri; a three-million-candlepower searchlight that was mounted on the chassis of a Cadillac; a Maxwell caterpillar tractor; and, the coup de grace, the Militor. Custom-built for the excursion at a price of $40,000, the Militor was described as a wrecker-winch or an "artillery wheeled tractor." Used many times throughout the trip, the Militor could be driven into the necessary position and then the winch bolted to its rear end could haul out any stranded vehicles. The Militor also contained a heavy iron bar that could be extended to the ground to provide the necessary anchor against which the winch could pull heavy loads.[86]

Just a few days on the road, after departing Bedford, the convoy arrived in Greensburg, Pennsylvania, sixty-three miles away and was met with the largest party yet. Although the town's population was approximately fifteen thousand, the crowd was estimated to be three times that size. Veterans of every branch of the military, including of the Civil War and Spanish-American War, marched with those who had just returned from Europe and then sat at long tables under the trees to enjoy a clambake and open bar that served lemonade (Prohibition had gone into effect ten days prior). Activities included boxing bouts, rifle shooting, and performances by musicians and magicians.

When the day ended with an afternoon storm, the Militor was pressed into action to rescue vehicles in the convoy. Heavy rain and slick roads had led to chaos of various sorts, including mechanical failure. One of the Garfords burned out its engine bearings when the crankcase opened up and all the oil drained out. A Dodge delivery truck lost control and caromed into one of the ambulance trailers. The Militor took all three vehicles under tow, but things quickly turned even worse. Lightning struck a tree fifty yards from a GM cargo truck, which skidded off the road and down a hillside and splintered apart. Eisenhower later reported that the truck was unrecoverable.

Despite this grim reality, Ostermann's boosterism of the convoy never skipped a beat. "Pennsylvania right now," he told the Daily Tribune, "doesn't need to take a back seat in the union in the matter of highways."[87]

The suburbs of Pittsburgh were more than ready to welcome the convoy later, on July 11. The city proper held "Transport Day," and the *Gazette Times* declared, "A Day of Vital Interest to Every Business Man and Firm in Pittsburgh. A Day Given to the Nation's Right Arm—The Motor Truck." The primary publicist, Colonel McCoy, never commented publicly on any of the mechanical challenges that the convoy had already faced. Two days later, though, the *Gazette Times* ran a cartoon with a group of men gathered around a car with a broken rear axle. "Human beings used to run without doctors," spoke one of the characters. "Wonder if cars'll ever run without mechanics?" Certainly, the convoy would not![88]

After he joined the convoy, Eisenhower's observer status required that he keep a daily report. In it, the young soldier singled out the Packard trucks for his admiration: he wrote that they encountered "very little difficulty." "These trucks surmounted the stiffest grades with motors running quietly and easily," he wrote. "One Packard truck was badly overloaded during the entire trip," which he stipulated as fifteen hundred pounds beyond its one-ton capacity. "The performance of these three trucks is considered remarkable."[89]

His youthful chagrin was caving as he watched the remarkable American ingenuity at work. In this moment of transition, there was simply no better place for a young American, like

Eisenhower, to be than moving as modern-day pioneers across America in a fleet of Packards and other cutting-edge machines. The First Transcontinental Motor Train was under way to blaze a path to the American future.

Now, if it could just locate some good roads on which to travel.

By the time they had reached the Midwest, participants in the convoy had noted one of the primary characteristics of their trek: dust and dirt. They proudly wrote "Gas Hounds" in chalk on the sides of their vehicles.

CHAPTER THREE:

Paving the Way

There was no real question about which path the convoy would take, particularly after they'd left Ohio. There weren't many options for what roads to take out west, and the route was strategic, part of a carefully planned publicity campaign. As it was originally conceived, the convoy consisted of two parts: one moving north and west along US Highway 30 and the other heading south to Atlanta, Georgia, before veering west. By 1919, the LH was taking shape. Although it was incomplete and parts of it were barely passable, a version of the "Coast-to-Coast Rock Highway" lay ready and waiting, with the LH. Even prior to the launch of the convoy, enthusiasm for a national roadway was growing. Following the eventual decision to operate the convoy as a single procession, the agenda of those shaping publicity efforts—particularly Joy and Ostermann—left no doubt that the convoy would follow the path of the LH whenever possible. Through this public demonstration of the value of the LH, then, communities might be convinced to invest in its completion.[90]

The most notable deviation from the LH came in eastern Ohio, because of an invitation from Harvey Firestone, the tire and rubber magnate who was partly underwriting the excursion.

Firestone wanted the crew to be feted at his Ohio estate, Harbel Manor. The Stars and Stripes film crew captured footage of the soldiers at Firestone's chicken dinner in their honor, and audiences would later see the celebration at cinemas throughout the United States. In his white linen suit, the tire magnate chatted with McClure and his officers, including Eisenhower. Eight speeches were intermingled with performances by a band and a singing trio. Mutt and Jeff, one of the army's traveling acts, was also filmed performing.[91]

The convoy rejoined the LH at Canton, Ohio, and as it moved westward, a routine, of sorts, took form and can be seen in the stop in Delphos, Ohio. First, a scout crew would arrive before the convoy, earlier in the day or the evening prior. McClure was usually part of this crew. They'd finalize logistical arrangements and interface with local media and community leaders as part of their booster efforts. In Delphos, after the scout crew had gone ahead, the first trucks arrived around 3:30 p.m. and were piloted through the community, passing the gathered crowds and ultimately arriving at their camp. As soon as camp was pitched, Delphos Mayor George N. Leasure delivered brief comments and presented McClure with the key to the city. He invited all the men to dinner at Saint John's Auditorium and sketched out activities hosted by the community that would follow the meal.[92]

"Everything free" was the town's slogan for the festivities, and, indeed, Delphos was good to its word. Two hundred men partook of dinner and leisure activities such as swimming; meals were sent to those who remained on duty with the vehicles and other equipment. Delphos, like other communities, made swimming pools available for soldiers. Typically, they also supplied swim trunks. In Indiana, though, the community didn't have a sufficient supply of trunks, so they erected a canvas screen around the pool and soldiers "could revert to the good old swimming hole costume with perfect propriety." At dark in Delphos,

They were entertained at many stops; this event in Ohio was
hosted by Harvey Firestone, the manufacture of tires.

the convoy's searchlight swung to life and flares and signal rock-
ets were launched along Main Street. Dance halls erupted to life
and remained active until late in the evening. The Knights of
Columbus provided the soldiers with postcards, writing paper,
and pens, and then mailed messages to friends and families. The
following day, the *Daily Herald* reported that the soldiers "had
conducted themselves as gentlemen." In typical military fashion,
trumpet reveille at 4:30 a.m. began the process of mobilization,
and the convoy rumbled to life shortly thereafter. And they were
on to the next stop, which undoubtedly waited for their arrival
with great anticipation.[93]

The soldiers could settle into a fairly standard script for each
of their visits. Although some demonstration of equipment took
place at each stop, military officials offered a celebratory appreciation for the World War I victory rather than a show of might. National pride was the central message of the convoy, at least from
a military standpoint. The idea that Americans could one day

make the same journey as the convoy would simply have been suggested by its presence. However, Dr. S. M. Johnson of the National Highway Association pressed his vision of transcontinental roadways along the away.

Johnson spun his comments to appeal specifically to the people at each stop. For instance, when they were in Indiana, he spoke about how his grandparents had traveled to Fort Wayne in 1832 at a clip of ten miles per day. Now, he stressed, you could go twice that distance in one hour. The nation was changing rapidly, he added, with more people traveling by car than by train, and with half a million trucks already on the road and swift promise for that to be multiplied tenfold. With this increase, he stated logically, we need good roads, and "the railway is no longer capable of meeting the transportation needs of the country." Certainly, support could be shown through applause, but Johnson also urged community members to pay a five-dollar membership fee to join the LHA and keep "the big transcontinental route in good shape." With membership, donors received the well-known LHA radiator emblem that decorated many vehicles throughout the nation. There was also an unspoken weight to Johnson's appeals: it was rumored that communities who did not properly celebrate the convoy risked losing their connection to the vital road. With the route in flux, the LHA's organizers could easily choose another path.[94]

Often speakers connected the convoy to larger policy initiatives beyond the Lincoln Highway. In the grandest vision, this roadway was just one in a new federal system that would be commissioned and funded through the Townsend Bill. Having been under discussion for years, the bill would finally be passed in 1921, two years after the convoy.

Business interests were also evident during the trip. As the truck train moved through the Chicago region, for instance, the *Star* reported, "The gas and oil men, the tire men, the auto supply men swooped down on [McClure] by the score, and there were newspaper men, moving picture operators, camera fiends, and good roads advocates, all anxious to speak with the man who was doing this big thing for the government." In response to the road-building boosterism, the paper noted that McClure stuck to his basic motto: "We're going to get there; that's all." That was no small feat in itself. The reality of managing these new machines on the road was not nearly as glamorous as the fanfare at each stop.[95]

More than anything, the convoy's progress suffered because of the emerging technology itself. With these new vehicles, the mechanical difficulties were an almost endless nemesis. Jackson blamed "bad gasoline supplies" for the repeated difficulty of starting vehicles at many stops. In Eisenhower's internal post-convoy report, he described mechanical problems as "slight." But describing the human errors that impacted the trip, he did not mince words:

> Reports of officers with the convoy indicated that the vehicles had not been properly tested and adjusted before starting the trip. This occasioned many short halts on the part of individual trucks, to adjust carburetters, clean spark plugs, adjust brake bands, to time motors and make minor repairs of this nature. It was evident though that many of these difficulties were caused by inefficient handling of the vehicle by the driver.
>
> As the trip went on, it soon developed that difficulties arose much more frequently in some types than in others.[96]

Eisenhower specifically criticized the Garford trucks, one of which had to be abandoned on the trip. In particular, chain-driven engines could not be operated in the sandy areas of the western United States. This meant that the heaviest trucks often had to be towed through extensive stretches, which slowed the convoy's progress considerably. Eisenhower also observed that the mix of different vehicles made the trip extremely difficult and that they shouldn't be mixed in future travel.

Eisenhower's report reveals his role as an observer. He writes about what he saw and facts about the vehicles, but he generally doesn't offer his opinion. In one qualitative comment, though, in arguing that military needs demanded good roads throughout the nation, he described the wide variability in road types that they had encountered in 1919, observing that

> the trucks operated very efficiently and easily on the smooth, level types, but that on rough roads, sandy ones, or on steep grades the truck train would have practically no value as a cargo carrier. The train operated so slowly in such places that in certain instances it was noted that portions of the train did not move for two hours. . . .
>
> It was further observed that in many places excellent roads were installed some years ago that have since received no attention whatever. Absence of any effort at maintenance has resulted in roads of such rough nature as to be very difficult of negotiating. In such cases it seems evident that a very small amount of money spent at the proper time would have kept the road in good condition.[97]

On the subject of personnel, Eisenhower noted that a lack of training among the enlisted participants slowed the truck train's efficiency. He urged that future groups should be properly prepared and that this would make caring for their needs simpler. Feeding convoy personnel also presented consistent difficulty. Although each community hosted meals for the soldiers, the truck train itself included "trailmobile kitchens" that were used for basic cooking needs. As Jackson explained in his report, these were "poorly constructed, and the ranges on both kitchens sheared their bolts and fell off the chassis on excellent paved roads. On poor roads the cans frequently bounced off, and it was practically impossible for a man to retain his seat on the vehicle." Also, the cooking range repeatedly fell off its trailer and required repair. In fact, neither of these two kitchen trailers completed the trip.[98]

Despite these obstacles, Eisenhower reported that the convoy "was well received at all points along the route." The activities at each stop left little doubt of the dual mandate of the excursion. Corporate, military, and government agendas—all were merged into a joint vision of a mobile nation that would benefit all economically *and* keep Americans safer. "It seemed," wrote Eisenhower, "that there was a great deal of sentiment for the improving of highways, and, from the standpoint of promoting this sentiment, the trip was an undoubted success."[99]

Jackson, as leader of the convoy, was happy to report how well the public received them: "The interest of the general public in the convoy was evidenced by a whole-hearted hospitality which never failed from beginning to end of the trip, and which was quite as spontaneous in the small towns as in the larger cities. Everybody was glad to see the trucks and the men, who were showered with every variety of refreshments all along the route." Particularly in Jackson's report, it is clear that in the aftermath of World War I, the convoy helped galvanize support for returning veterans, while at the same time guiding people to look towards the nation's future.[100]

LEAVING THE mid-Atlantic and crossing into the Midwest, the convoy's next most publicized stop was in Clinton, Iowa, where—like in many other places—the community attempted to use this event to leverage modernization.

Clinton was unlike most of the farming towns one would find across the Iowa landscape during this time. Although its

Ike and three others in Ohio.

heyday in the 1890s had come and gone, the town's diverse economy was industrial, largely based in lumber. Dozens of sawmills cut trees that were brought down the river from the Great Lakes region. Planks from Clinton were used for building houses, barns, and fences throughout the state and region until 1897, when all but two of the mills closed. Many Clinton residents moved onward, deeper into the West in search of emerging opportunities.[101]

Late in the afternoon of Tuesday, July 22, in a cloud of dust, the convoy crossed the Mississippi River on the High Bridge between Fulton, Illinois, and Lyons on its way to overnight and enter Clinton. Erected in 1891, the rickety bridge, like most passages in the state, had been built to carry wagons. Its loose planks and unstable condition made the passage of a wagon unnerving, let alone that of a five-ton truck. Subdividing into three or four vehicles at a time, the convoy crept over the unsteady bridge and arrived safely at its camp site in River Front Park. Ostermann and Lieutenant Doron had arrived the night before in their red, white, and blue Packard Twin Six.[102]

Even though the town's population was no longer growing, as many as twenty thousand residents turned out to meet the convoy. Spectators sat by the banks of the river and were treated to exciting displays, including machine-gun firing and roaring high-powered engines. Members of the crew proceeded to the local YMCA, where they rinsed off the dust of days on the road and then enjoyed an evening of Clinton's hospitality, including a band playing at the park and a jazz orchestra at the Coliseum. In this transitional moment in the development of the West's economy, residents hoped that the convoy might lead the way to the future—any future. Auto travel seemed as if it might just be the key.

Like the High Bridge, the dirt roads of Iowa were remnants of the past, designed to be used in a very different fashion than for the convoy. Two weeks without rain thus far made daily passage a swirl of thick dust. "An' the hardest thing about it," said one sergeant who had fought in the Argonne, "is to see the sweet, clean faces of the tourin' car girls smile up at us and not be able to smile back."[103]

Despite the conditions, after sixteen days the convoy was already a success: it had gone 906 miles—already farther than any previous truck train. They would continue, with the nation watching. And the dusty roads actually offered an opportunity, one arranged by design. Traveling through deplorable thoroughfares, the convoy could demonstrate the need for investment in better roads. The going was not supposed to be easy and, indeed, it would not be.

The convoy headed toward Cedar Rapids and was subject to the vagaries of such travel: when the conditions were dry, severe dust stirred with each passing vehicle, filling human lungs and corrupting the airflow for vehicle engines; when it rained, endless and overwhelming mud slowed passage and often sucked in wheels, requiring time-consuming extraction.

In Iowa, paving the roads was an active, highly debated issue: fourteen counties favored the public expense, and four held out—said by critics to be in the "mud column"—arguing that such activity marked dangerous federal overreach. But even the farmers of Iowa who could see no neighbors from their homes knew that effective roads would make trading their produce much simpler.[104]

In this moment of transition, American communities were assessing their best path forward into a new era. Eisenhower and his fellow travelers in the convoy knew the paving binge couldn't happen soon enough to help their passage, but their efforts were part of a massive push to awaken Americans to the fact that such modernization efforts were worth the cost.

FROM THE early discussions of boosters in 1911 and 1912, there was little doubting the benefit of a transcontinental road. There was, however, no public expectation that the government should or would finance such an infrastructure project. These boosters had to drum up funds, so they approached communities directly. But the automobile industry also played a key part. Fisher estimated that the road would cost $10 million, and he sought to raise the funds privately from auto companies, asking that they pledge one-third of one percent of their gross revenue. Between 1913 and 1915, he received strong support from automobile executives—except Henry Ford, who had no interest in contributing.[105]

Most importantly, Henry Joy became involved and helped to center the road-building initiative out of his hometown of Detroit. Passion for his first few Packards drove him to purchase the company, and in 1912 Packard threw its backing behind the highway project. As the group began to consider how they could broaden the appeal of the LH, they determined that the name "Coast-to-Coast Rock Highway" lacked the necessary glamour and appeal. They actively considered "American Road" until, at an April 14, 1913, meeting in Joy's office, Fisher proposed "Lincoln Memorial Highway." By late June, they had incorporated the LHA and based it in Detroit—with Joy as the first president.[106]

As the planners of the highway laid out the preferred route, offers of interest poured in from western cities. These local leaders thought the LH was akin to the transcontinental railroad and wanted to make sure that their city was not bypassed by a new era of transportation. In August 1913, Joy set out to quell the flood of requests when he announced: "The arguments of all interests have been and are being weighed." However, he said, that they could not all be routed. Such a road, he explained, "would indeed be a devious and winding journey in this great America of ours. . . . The decision must be confined to one permanent road across the country to be constructed *first*, no matter how desirable others may be and actually are." For the financing, Joy tapped his network of "idealists" who each led companies involved in making the raw materials of automobility: autos, tires, and cement. They referred to the Lincoln Highway Project as their great patriotic work, despite the fact that financially they each stood to benefit mightily.[107]

The LHA published the official route on September 10, 1913, and Joy addressed governors in "The Appeal to Patriots," 150,000 copies of which were sent out to newspapers and car dealers nationwide. Pennants were flown alongside American flags at participating showrooms, with banners proclaiming, "Without dreams the world could accomplish little!" The project was repeatedly compared to the construction of the Panama Canal, which was underway in the southern hemisphere. Across the United States, events were held to dedicate the route of the LH on October 31, 1913, including bonfires and fireworks in hundreds of cities in the thirteen states along the planned road. There were concerts and parades; preparations were made for dancing on the highway.[108]

In his speech at the dedication ceremony in Jefferson State, Iowa, State Engineer Thomas H. MacDonald captured the vision of many Americans for the LH. He forecasted that it would quickly become "the first outlet for the road building energies of this community." The LH would spur the construction of radial routes and eventually "great transcontinental highways." The system of radial connections should be the emphasis for communities, he urged. For those joining the system, he foresaw that they would ultimately tie into "one great continuous road across the state."[109]

In a model of private-public cooperation, "The Seedling Mile" program allowed the LHA to assist paving projects along the route between 1914 and 1919. The Portland Cement Association agreed to donate one mile of free cement to any town that provided the necessary grading and continued maintenance afterward. This venture fueled progress in paving; however, the lack of coordination also contributed to a randomness that undercut any hope of establishing public confidence in road conditions overall. Politicians, boosters, and military officials hoped the convoy might be the perfect impetus to unify purpose and confidence and to launch the nation into the new era of the automobile.[110]

The challenges of meeting Joy's original goal of completion in 1915 immediately became evident. To garner support, the LH became one of the most extensive applications of boosterism in the auto industry. Efforts to package the LH took many forms, including guidebooks, documentary films, lapel buttons, pennants, radiator badges, stickers, and stationery. While there were clear economic similarities to the potential realized by the railroad, the highway (which eventually became known as Interstate Route 66) promised an era of individualized consumption that was tied inextricably to the automobile. The LH presented Americans with the first attainable expression of the new frontier of automobile tourism. Historian Richard Weingroff notes, "During the early years, a trip from the Atlantic to the Pacific on the LH was, according to the LHA's 1916 *Official Road Guide*, 'something of a sporting proposition.'" Estimating a driver speed of eighteen miles per hour, the LHA estimated the trip would take twenty to thirty days.[111]

The guide also offered drivers instructions for how best to weather the challenges of this early travel. As infrastructure was scarce along the road, Weingroff writes, the guide encouraged motorists to "buy gasoline at every opportunity, no matter how little had been used since the last purchase." When faced with standing water, motorists were encouraged to first wade through prior to driving their vehicle in. Although the guide did not advise drivers to carry firearms, full camping equipment was considered essential, particularly west of Omaha, Nebraska. According to Weingroff, the guide advised that the equipment needed by drivers included "chains, a shovel (medium size), axe, jacks, tire casings and inner tubes, a set of tools, and, of course, 1 pair of Lincoln Highway Pennants." Finally, with an eye toward the mud proliferating in most roadways, the guide advised travelers simply, "Don't wear new shoes."[112]

The LH effort demonstrates the growing cultural interest in transitioning to auto travel that set the stage for the largest booster action of all: the 1919 convoy. Indeed, the Packard that Joy drove during the convoy was the LH model!

However, at least one additional, crucial dimension pressed the national interest in the convoy: military applications. In terms of logic, roads were an essential portion of military preparedness, particularly in this new energy transition. And the outbreak of World War I underscored this need—particularly a little-remembered invasion of the American mainland by a foreign power.

As early as 1915, Joy argued publicly that "national pre-paredness" in this new era required a commitment to motor vehicles as well as a functional system of accessible roads. As global war threatened in 1915, roads were important enough that twenty-five cadets from the Northwestern Military Academy of Lake Geneva, Wisconsin, were sent the LH's intended route to San Francisco. Their mini-caravan included an armored machine-gun car, an armored wireless station, two "balloon destroyers," an ambulance, and three cars. The expedition's commander, Colonel R. P. Davison, reported afterward, "As a military necessity, the LH should be constructed so that the heaviest artillery could be rushed from one coast to the other with the rapidity and efficiency which German roads have al-lowed in Teuton maneuvers."[113]

Soon afterward, in March 1916, Pancho Villa and his raiders crossed the border into Columbus, New Mexico, on behalf of Mexico. The raiders burned parts of the town and killed seven American soldiers and eight civilians. President Woodrow Wil-son ordered a "Punitive Expedition" to pursue Villa into Mexico. Packard received the Southern Department of the Army's order two days later: two motor companies of twenty-seven trucks each. The following afternoon, the order was complete, and the trucks were loaded onto a special train bound for El Paso. From that point, they drove more than six hundred miles into Mexico and completed a very successful mission. Although history re-members the 1917 "Zimmerman Telegram" involving Mexico as a catalyst for American involvement in World War I, the border was already becoming contested terrain before that point.[114]

General John J. Pershing dubbed the Packard trucks the "Fly-ing Squadron," despite the fact that they often had to be hauled by hand through mud and sand. In the words of the Quartermas-ter Corps's historian, "Any lingering doubts as to [motor trans-port's] superiority over animal-drawn transportation were forever settled [after the battle]." As the call for involvement in the war grew during 1916, the *New York Times* reported, "Preparedness Means Millions for Motors." Wilson ordered the United States into the war in April 1917, and within eighteen months, the army owned eighty thousand trucks—many of them Packards. A new urgency had emerged for a road across America, one that grew out of national security concerns.[115]

Throughout 1917 to 1919, the federal mobilization for the war effort required an unparalleled movement of men and materiel, particularly throughout the northeastern United States, ending at various ports for transshipment to Europe. The government took over the railroad system; however, it was soon overwhelmed. Trucking on roadways became an absolute necessity under the oversight of the newly formed Highways Transport Committee, and, whenever possible, the LH provided the primary route. It was the nation's best road and was now demonstrating its worth.

While leading a convoy of materiel from Detroit to Baltimore during World War I, Ostermann hatched an idea that combined his mission with his passion: once the war was over, why not send a similar military convoy across the entire length of their American road? He discussed the idea with military representatives and found a taker in Captain Bernard McMahon, who laid out the initial plans and preparations. With the war in the nation's past, such a trek could link together the disparate local road-building initiatives while also celebrating a victorious military and a new national beginning. After these initial efforts, McMahon turned this project—the Transcontinental Motor Train—over to McClure, though he still accompanied the truck train for its entire trip. When Ike and the rest of the convoy tore through Iowa in 1919, it had clearly become a joint effort between military and industrial representatives.[116]

*   *   *

FROM HIS early days in Abilene, Ike knew what it was like to live in remote segments of the Midwest. His choices, of course, had brought him into the bucket seat of one of these great vehicles. Now, as the convoy proceeded west through Iowa on the LH, small communities, like his hometown, derived the benefit of one of the essential commitments of the road: unlike highways of today, including US Highway 30, the LH did not bypass towns and, instead, passed from one Main Street to the next.

And, so, the convoy did as well—almost entirely. One of the first departures from the LH came shortly after departing Pittsburgh, and it clearly reveals some of the mixed motives at the root of the convoy. The primary force behind the convoy was boosterism, so influence meant everything.

In the twenty-first century, most of us happily take roads for granted. We expect them to exist where we need to go. And we also expect them to be at least moderately passable. Many of us probably phone the local hotline to report potholes or any significant bump in the smooth asphalt surface. These expectations represent our own moment in energy use; the willingness of drivers and riders to tolerate absolute unreliability in their roads in 1919 grew out of their own cultural moment. The shift in perspectives toward a demand for reliability began with the convoy—particularly as it passed out of Clinton, Iowa.

A similar sentiment for good roads had been seen as well a few days earlier when the convoy departed DeKalb, Illinois. The local *Daily Chronicle* reported that the nation's Good Roads Movement was underway: "People the state over, regardless of whether they own a car or not, are beginning to see the benefits possible from good roads." And to nationwide travel on good roads, the LH needed to be improved. At present it was "full of detours . . . in such condition that no car could stand up under such wear and tear. . . . Think of the pleasure that will be possible when the cement road has been built from cost to coast."[117]

In 1919, though, throughout the Midwest, they had a long way to go. McClure reported to Washington that the ninety miles of road through this area in Illinois was little more than a dirt track in "deplorable" condition. If dry, the dirt roadway section was overwhelmed with dust or if wet, teeming with unforgiving mud. Covering 906 miles in its first sixteen days, the convoy had already bested any previous trucking effort; however, the conditions of the road westward made participants now long for those earlier stretches.[118]

With Iowa's agriculturally oriented layout, the state's population was spread out. Most of these rural inhabitants shared the opinion that rock roads (normally a large gravel) were a luxury of the wealthy urban elite—"peacock alleys," they called them. During a 1915 drive through Iowa, Henry Joy had written indignantly that in the state, "Not a wheel turns, every farm is isolated. Social intercourse ceases. School attendance is impossible. Transportation is at a standstill. Millions of dollars' worth of wheeled vehicles become . . . worthless." Iowans had passed a 1919 law that required road improvements; however, counties had to vote in favor of raising the necessary funding. By the time the convoy passed through, seventeen counties had voted in favor of road improvement and those that had not were deemed members of the "mud column."[119]

Clinton was a unique Iowa town, given its industrial past. Unlike many Iowan towns, it had improved roads. When its mills had begun to experience economic downturn, the town had temporarily propped the industry up by surfacing its roadways in creosote-treated wood blocks. This was not a desirable material—during summer months, the streets both buckled and stank; however, the roads were not merely dirt. And dirt—particularly wet with rain that turned it to mud—is what the convoy met as they continued the trek through Iowa.[120]

West from Clinton, towns such as Cedar Rapids and Marion feverishly sought advantage of any sort over neighboring
farm towns and cities. These two towns, for instance, had bat-
tled over roadways since the LH was first laid out. Marion won
the highway but required an otherwise unnecessary northern
tack. Although the convoy traveled through Marion, leaders of
Cedar Rapids realized what needed to happen for the good of
their town's future: they paved their road east toward Mount
Vernon in the ensuing years. In short order, Cedar Rapids
received most of the traffic, and soon after, Marion lost its status
as county seat and even the LH. Even in small-town Iowa, the
roadways to the future were paved.[121]

HOW COULD the entire sprawling nation update its roads
simultaneously? Not since the Roman Empire had the world had
experience with such efforts at widespread infrastructure modernization. Coming out of World War I, this was a basic quan-
dary besetting the nation.

Bricks had provided the material for the first "hard rock"
roads in some parts of the United States. In Ohio, for instance,
bricks made from high-quality shale and clay had been used in
building for a century. The first bricks were used on a roadway
in Steubenville in 1884, and Canton soon began mass-producing
paving bricks. During the next few decades, the town's Metro-
politan Brick Company became America's largest manufacturer,
supplying raw material for roads throughout the nation.[122]

Energy transitions are full of temporary outtakes—paths not
taken—and bricks are one for this moment. Hard roads were
necessary for auto travel; however, bricks were not a practical
solution. It is estimated that a mile of road, twenty-five feet wide,
required half a million bricks. While bricks could be locked
into building construction with cement, their use for roads was

flawed—particularly when the roads were used consistently by heavier vehicles.[123]

As one symbol of the lessons of this transition, Canton's brick industry slowed as the material for roadways moved to concrete and, later, to asphalt and blacktop.[124]

A hodgepodge of initiatives, funding measures, and booster efforts made American roads of the 1910s a crazy quilt of different materials and stages of development—even the LH. All over the country in the 1910s, Good Roads Days drew volunteers to work on transforming the section of the LH that passed through their community into a hardened version. For instance, in passing through Illinois over two days, the convoy covered 172 miles of road, over dirt, gravel, brick, and concrete. Still, wherever it passed the convoy of 1919 kicked up a cloud of dust in most parts of the United States.[125]

The variety of options for road coverings fed the competitive nature of developing communities in the Midwest and West. Similar to bandwidth in the internet era, towns wanted to make the choice that would attract further development and usage. In Iowa, as towns such as Cedar Rapids and Davenport considered how best to situate themselves for future travelers, step one appeared clear to the *Davenport Democrat* when it wrote, "The fact that automobile travel will follow the hard roads, whether we name them Lincoln Highway or what not, makes it plain that Davenport will be the principal entry to the state from the east if we are prompt in offering a hard road across the state from this point." A bit farther west, townsfolk in Ames announced that the convoy provided proof of how dirty travel was, whether muddy with rain or dusty with dryness. "Get Iowa out of the Mud!" called out local boosters.[126]

In Des Moines, the *News Republican* wrote critically about the embarrassing state of their roads for the convoy's journey. A reporter observed, "The Lincoln Highway will follow the line of 'least resistance'—a paved highway is that line, a mud road is not. . . . Are we, living on the Lincoln Highway—the greatest asset in Iowa—to remain asleep at the switch? . . . It is time there was concerted action all along this highway, to save it from the junk pile. Will we take it?"[127]

The issue of improving roadways was also being actively debated in Jefferson, Iowa, as the convoy approached in July 1919. With the vote on road improvement scheduled for July 28, two days after the convoy came through town, the topic of paving was a frequent subject in the local *Jefferson Bee*. In the July 9 edition, the *Bee* ran a photo of a car stuck in mud with the caption, "A familiar sight last winter and spring. . . . The only permanent solution is paving." With a positive vote, the editors forecasted that sixty-five miles of paved road would be added in the county within five years, and the bill would be paid with federal aid from the 1916 Highway Act. On July 25, the *Bee* quoted the report of board supervisors who had traveled to neighboring counties with concrete on roads: "We went away with open minds . . . and came home convinced that concrete roads . . . are practicable, durable, and a good investment."[128]

These sentiments were soon reinforced by the convoy's speakers on July 28. In particular, Dr. Johnson stressed that the auto and truck industry was growing explosively without a proper infrastructure in which to use it. Unlike railroads, he explained, there was no centralized industry to take responsibility for the appearance of adequate roads. He argued that while some farmers might think that they could not afford the tax dollars to construct roads, once they owned vehicles, bad roads would cost them more in gasoline and wear and tear. After all the fanfare around the roads, it must have come as no surprise when two days after the convoy's departure, Greene County approved over $1 million for road paving. The country voted three to one in favor, which was the highest proportion of favorable votes in the state. Clearly, the spectacle of the convoy had influenced voters.[129]

*   *   *

WITH EACH passing mile of the unrefined western roads, the dirty, worn state of the convoy's vehicles and human participants became the single best representation of the need for concrete roads. As they neared the end of Iowa, McClure, with all his boosterism, could not ignore the elephant in the room as he kept up his report: "The dust was there, but we don't mind that if the road is good." He added that the convoy "is doing more for good roads than anything else in the history of the country. You will see more good roads in process of construction after this trip than you ever had any idea of." Indeed, by the mid-1920s, the LH was paved all the way across Iowa.[130]

Crossing Nebraska, the LH stretched five hundred miles at a slow, nearly imperceptible ascent, with very few towns and almost no paved road. Indeed, the road conditions were even worse than those encountered in Iowa. The eighty-two miles from Grand Island through Kearney to Lexington took eleven hours, particularly when one of the trucks sunk into a ditch of soft dirt up to its axles—about four feet deep. The Militor took two hours to haul the truck back onto the stable road after they had first dug two feet into the roadbed. Soon, the Militor was also required to tow another vehicle the last dozen miles of the day. Its engine had burned out under the strain of the muddy roads.[131]

On August 1, in response to the approaching convoy, H. D. Watson hung a sign outside of his Kearney farm that announced it as the halfway point between coasts, 1,733 miles from the major urban area on each coast. A few days later, the muds of Dawson County repeatedly mired almost every vehicle other than the Militor. The routine grew so tiring and repetitive that Jackson and McClure stopped keeping track of each mishap. Even the engines of the motorcycles clogged with the thick mud. Dr. Johnson's telegram to Secretary of War Baker declared that "for the first time since war was declared, the army has been forced to surrender—to Dawson County Mud." Embarrassed, the local newspaper wrote, "Really folks, it's a shame. . . . There is no good excuse for tolerating longer the rotten roads that we know we have."[132]

"Crusted watersoaked quicksand" is how McClure described the road fourteen miles west of North Platte on August 5. While it may have seemed unthinkable that the road conditions could worsen from those seen in previous days, this rain-soaked stretch proved even more problematic, particularly because to the soldiers' eyes the conditions appeared stable. However, once one set foot—or especially wheel—on the surface, it gave way and the vehicles sunk to their axles. Only a few of the lighter vehicles passed through, and the Militor and tractor were left to excise every other vehicle. A mere two hundred yards' passage wound up taking the convoy over seven hours. Entire bridges needed to be rebuilt and planks laid to corduroy the most problematic stretches. The swamp of this section of Nebraska had cost the

Convoy truck stuck in mud being pushed by crew members.

Particularly in long stretches of the American West, little effort had been devoted to making this territory accessible to vehicle travel. Therefore, moments such as this made the convoy's vehicles a demonstration to communities and voters that infrastructure was crucial to the nation's transportation future.

travelers an entire day. "Entire personnel quite exhausted by se-
verity of seventeen hours continuous duty," reported McClure. [133]

Eventually, they entered the stretch of the United States
referred to as "The Great American Desert" by Lieutenant
Stephen H. Long in 1820. McClure simply described it as
"sparsely settled" in his official report. Learning from their
recent experience and in awe of these new surroundings, the
convoy leaders set more practical goals for its daily passage and
established a routine for best managing the vehicles through the
poor route. In fact, the worst was behind them. The final stop
in Nebraska came in Kimball, where the soldiers washed in irri-
gation ditches and enjoyed the town's festivities. The following
day, the convoy crossed out of the state and began the final leg of
the journey. [134]

WITH LESS mud, they were soon able to travel at a faster clip,
even though the terrain remained challenging. For the soldiers in
the company, many of whom had just returned from the western
front of World War I, the excruciating difficulty and plodding of
the convoy's passage was business as usual—and certainly a vast
improvement over doing so while being under fire. Young Eisenhower's recollections reflect this discipline when he recalls simply, "We were kept busy taking care of breakdowns and trying to
improve our march discipline. But it wasn't all work and it wasn't
all discipline." Eisenhower was clearly having fun. [135]

He was not an ordinary member of the company, yet he
seemed to rejoice in being among "regulars." Eisenhower was a
young man coming of age, and already involved in big things,
even if he yearned for more responsibility. By jumping into the
convoy experience, Ike can serve as a model for young people in
any era. He identified emerging aspects of his moment in time
and sought opportunities to participate. With this kind of focus,

Ike and Brett created many adventures for themselves and others.

a future leader can learn firsthand about the texture of the nation and its populace. We can see this easily in the context of hindsight, and yet the benefits that Ike derived on this journey were largely unintentional. Like young people in 1919 *and* in 2024, of any generation, more than anything Ike was open to life as it unfolded before him. He made his choice of a path toward the future when he hopped on the convoy, though he was not aware of the impact it would later have. Speaking of the veterans of World War I, he later wrote, "We might have filled the time at the card table or the bar. But there were chances to reach out, to search for duty that was more than perfunctory."[136]

From his own perch of hindsight, Eisenhower focused most on a young traveler's opportunity to see the nation and meet its citizens. He came to later cherish the "nodding acquaintance" that he acquired with people and places "across the east-west axis of the country" during the trek. He felt that the convoy's process made this interaction possible in a unique way:

We were always, of course, routed through the main streets of each community. Our snail's pace enabled me to observe anything different or unusual. At every overnight stop where there was a town, we were welcomed by a committee and if only to demonstrate that I was not solely an Army propagandist, I tried to learn as much as I could about local interests. Much that I learned was quickly forgotten. But enough stayed with me so that decades later, it had its uses.[137]

Most of Eisenhower's personal recollections stem from the enjoyment he took with the camaraderie among the soldiers. As the convoy moved deeper into the western United States, Eisenhower's regional roots took over, and he found ways to offer value, particularly to soldiers from the East, through his knowledge of the area and his sense of humor.

Eisenhower and Major Sereno Brett functioned much as a comedy team. One of their most hilarious pranks starred Ike as the fastest gun in the West. While driving through the plains of Nebraska one day, the two soldiers encountered many jackrabbits and were able to bag one with a shot from the .22-caliber rifle. Brett quickly hatched a plan and carefully positioned the rabbit corpse at the edge of the scrub for later use. "From a distance," recalled Eisenhower, "it looked fairly natural." When the two soldiers returned that evening with a few of their "eastern friends," Brett extolled Eisenhower as one of the best shots in the West—exploiting the soldiers' expectations of westerners and placing the young soldier in the lexicon of the nation's great gunslingers. Armed only with .45-caliber pistols in the last moments of sunlight, Brett suddenly burst out, "Stop! Stop! Look over there—see that rabbit!" Eisenhower takes over the story from there:

> At that distance, and in dimming light, one could see him only in imagination. We described him so carefully that the

easterners admitted they could make out his outline. . . .
[Brett] said, . . . "Ike, why don't you take a crack at him?"

I carefully aimed the pistol in the general diction of the
North Pole and fired. [Brett] exclaimed, "You've got him!
You've got him! He fell."

Never having seen the rabbit in the first place, the easterners agreed with Brett that it was surely a fine shot and they
marveled at such skill.

I sat with them while [Brett] raced out to pick up the big
jack. Held by his ears, the rabbit must have been almost two
feet long.[138]

In another ongoing hoax, Eisenhower and others convinced
the younger soldiers that Brett suffered from PTSD, and each
night, he demanded to pitch his bedroll off-site from the others.
Often, Brett fleshed out the joke with outbursts and whoops. In
Wyoming, though, his private camp was on a bluff 150 feet above
the camp. Brett spent the evening moaning and howling like a
coyote to unnerve the easterners. "It wasn't too good an imitation
for one who knew the real thing," recalled Eisenhower, but it was
obviously convincing to the soldiers from the East.

Further into Wyoming, Brett went to the local diner and,
after befriending some of the locals, staged another effort to
exploit stereotypes of the West. Just as he had coached, the locals described the community's current difficulties with Native
Americans living nearby. Once the eastern soldiers sat nearby
and were sure to overhear, the locals worried that an outbreak of
violence was imminent. All concerned established that a night
watch should be organized and the eastern soldiers drew first
duty. The ruse grew more elaborate as Brett borrowed a shotgun from a local and removed its shot. Finally, in the middle
of the night, he called out from the dark "in the manner of car-
nival Indians." Eisenhower explains, "The recruit took his

duties seriously, marching at attention around the camp as if on parade. Brett and several others spread out and from a distance would let out an occasional short yelp. Finally, just as we hoped, the sentry let go with both barrels—to arouse the camp, he explained later."[139]

Brett and Ike, comedy team, though, had come to that crushing, familiar moment when pranks begin to go bad.

It turns out that that soldier who had been the butt of their joke was assigned to send daily reports back to Washington. Upon learning that his log reported the first regional flashes of violence with Native peoples in three decades:

> Faster than any vehicle in the convoy, we shot off in all directions to find the man who was carrying that message to the telegraph office. We found him, took the story to the commanding officer, and pointed out that if such news reached the Adjutant General, he was unlikely to understand our brand of humor. The commanding officer went along with the gag, crossed out the Indian part of the telegram, and sent the rest of it off.[140]

Brett and Eisenhower even managed to keep the soldier who wrote the report from discovering that any of this had transpired. Which brings up a natural question: did these two pranksters *ever* own up? Fortunately for us, Eisenhower was wired as an old-school public servant, so he discloses the answer before we ask the question, when he recalls,

> From the beginning, we had expected that the fellow . . . would realize what was up. We had always intended to tell him long before we reached San Francisco. Instead we found that he had taken the whole thing seriously. Any attempt to explain it would humiliate him. . . . I am sure that in the almost half-century since, he has had no inkling that what he

and one or two others went through on that journey was as part of an audience for a troupe of traveling clowns.[141]

Not quite the same as the moment when George Washington admitted to chopping down the cherry tree, but a rare view of honest humanity from a leader who would become an American icon. Refreshing is one way of describing it.

Clearly, though, the truck train also made it obvious that Eisenhower—not yet a general or a president—was also not quite a normal soldier. The young Eisenhower worked hard, took firm stands to support basic structure and bureaucratic responsibilities, and firmly led soldiers who were often his senior in age. In hindsight, he described that the convoy had been "literally thrown together and there was little discernable control."[142]

For Eisenhower, the convoy sometimes—as with those shenanigans about the West—seemed a bit of a man's recreational outing. At other times, he had a different role, receiving treatment likely not offered to any other participant. In Boone, Iowa, for instance, Mamie Geneva Doud was celebrated as a favorite daughter. She had grown up there, and her father became a well-known millionaire in the West when the family moved to Denver, Colorado. When the convoy neared Boone, the townsfolk knew it included Mamie's husband and the father of the baby Doud Dwight, who had spent parts of his first year living in Boone with his aunt and uncle, Joel and Eda Carlson, while his father moved from camp to camp.

Although the stop in Boone would only last an hour, Eisenhower and Brett arrived ahead of the truck train and were interviewed by the *News Republican*. "I can't say too much about the condition of the Lincoln Highway. Imagine a great truck convoy of this kind out over 1,200 miles, practically ahead of its schedule!" Eisenhower exclaimed in a moment of seemingly deft political-speak. In 1919, the interview was a bit of color to a rather dull, drawn-out event; found a century later, the utterance of the future general and president marks a golden nugget for the historical archivist![143]

A bit farther down the road came the real highlight for Eisenhower: a visit from the wife and son, both of whom he hadn't seen for eight months. With her parents, Mamie and Doud Dwight (referred to as Dewie) met the convoy in South Platte, Nebraska, and traveled with it until Laramie, Wyoming. Mamie and her father had been charting the convoy's course from their home in Denver and had chosen the best spot to meet up with it. As they watched the trucks roll into town, she could hear Eisenhower hanging out the window of a staff car "hollering like a kid on a rollercoaster." Pressed for months by the strain of his service, Eisenhower came away from the "fine interlude" vowing to himself "that it would be nice, being in the West already, to apply for a leave with my family at the end of the tour—if indeed we ever reached the end."[144]

San Francisco, CA: Parades such as this one in Salt Lake City, Utah, welcomed the convoy at each urban stop. For the festivities in San Francisco, the convoy was referred to as the "land fleet" or "land armada," and with the US Navy's Pacific Fleet it formed a symbol of the ascendant United States on the global stage.

CHAPTER FOUR:

Completing the "Land Armada"

There was clear irony to the fact that after driving three thousand miles across North America, the convoy's conclusion required the vehicles to be still for the final stretch, jammed unceremoniously onto two ferries that sliced through the quiet San Francisco morning. The harbor area was instrumental to the staging of the final celebratory moments of the truck train.[145]

On the morning of September 6, the ferries carried the convoy from Oakland Harbor to San Francisco, passing six of the Pacific Fleet's new destroyers moored nearby. Two of the great ships lifted their anchors and escorted the ferries across the bay. Unloading at the ferry building, the convoy lined up to make the final leg of its historic trek.

Throughout the trip, McClure had filed his daily reports back to Washington out of duty and responsibility; now, though, these reports had become essential to coordinating a grand finale. Behind the scenes, officials were planning a feat of magnificent symbolism. Weeks earlier, the new Pacific Fleet of the US Navy had passed through the Panama Canal, and military leadership at the highest level had engineered timetables so that they would be in San Francisco when the convoy arrived. Despite each breakdown

Convoy cyclists guided the party and often rode ahead
to prepare the route.

and dour setback captured in McClure's reports from Wyoming
and Nevada, the desire to create a proper symbol for a nation ascendant predominated the convoy's final moments.

WHILE THE truck train would symbolize a changing world, the
naval fleet at the end of World War I already did. The Asiatic
Squadron and the Pacific Squadron were combined in 1907, and,
starting in 1910, a separate Asiatic Fleet maintained and secured
America's presence in this increasingly active area of trade and
expansion.

Over the ensuing decade, the Pacific force had become even
more essential. During the 1910s, the energy transition that would
impact American personal transportation and drive the convoy

after World War I began first at sea by powering ships. Seeking a strategic advantage of any type, Britain was first to entertain the idea of powering its navy with neither wind nor coal. Just as young Winston Churchill had altered land warfare by incorporating the tank, as discussed above, he also proved instrumental in the conduct of war by sea.

Like Eisenhower years later, Churchill was one of the political figures who identified an active energy transition and then openly maneuvered and manipulated it to his nation's strategic advantage. Powering the British Navy marked such strategic application of crude. While Churchill did not originally favor naval expansion, he clearly saw the advantages of powering ships with petroleum, including speed capabilities, flexibility of storage and supply, the ability to refuel at sea, etc. To express his strong support, he later wrote,

> As a coal ship used up her coal, increasingly large numbers of men had to be taken, if necessary from the guns, to shovel the coal from remote and inconvenient bunkers to bunkers nearer to the furnaces or to the furnaces themselves, thus weakening the fighting efficiency of the ship perhaps at the most critical moment in the battle. . . . The use of oil made it possible in every type of vessel to have more gun-power and more speed for less size or less cost.[146]

By 1912, Britain had formulated policies that would use petroleum to enhance its superlative Navy. As Churchill wrote, "the supreme ships of the Navy, on which our life depended, were fed by oil and could only be fed by oil." Churchill and Britain's military strategists focused on the great benefits for their naval superiority if they moved beyond coal; however, their decision also marked a defining moment in a new era of the culture of petroleum. Committing their fleet to petroleum meant that it became one of the most important commodities on earth—nations' security depended on it. And any nation wishing to compete with Britain had to follow suit.[147]

Two years later, when he addressed the House of Commons on June 17, 1914, Churchill's vision was clear:

> This afternoon we have to deal, not with the policy of building oil-driven ships or of using oil as an ancillary fuel in coal-driven ships. . . . Look out upon the wide expanse of the oil regions of the world. Two gigantic corporations—one in either hemisphere—stand out predominantly. In the New World there is the Standard Oil. . . . In the Old World the great combination of the Shell and the Royal Dutch. . . .
>
> For many years, it has been the policy of the Foreign Office, the Admiralty, and the Indian Government to preserve the independent British oil interests of the Persian oil-field, to help that field to develop as well as we could and, above all, to prevent it being swallowed by [others]. . . .
>
> [Over] the last two or three years, in consequence of these new uses which have been found for this oil. . . . There is a world shortage of an article which the world has only lately begun to see is required for certain special purposes. That is the reason why prices have gone up, and not because [of] evilly-disposed gentlemen of the Hebraic persuasion.[148]

So, on the eve of World War I, the status of crude changed dramatically, and the war would press its new importance on a global stage almost immediately.

For the US Navy the transition was not as rapid, but when it arrived in the San Francisco Harbor on September 1, 1919, the Pacific Fleet was undergoing further definition as a clear reflection of an emerging worldview after World War I. Just a few years later, in 1922, the Pacific and Atlantic fleets were combined to form the United States Fleet, which positioned a main body of ships in the Pacific and only a scouting fleet in the Atlantic. For the first time, the major weight of American sea power was assigned to the Pacific. With intensifying conflict during the 1930s, the US Navy again split in two and on February 1, 1941, the Pacific Fleet established its new headquarters at Pearl Harbor. Nine months later, of course, Japanese warplanes attacked ships and installations at Pearl Harbor and elsewhere on Oahu without warning, thrusting America into World War II. In 1919, though, the symbolism of the fleet's presence was the thing.

AS THE ferries carried the convoy's vehicles and personnel across the harbor on September 6, 1919, six of the naval destroyers were moored nearby. Two destroyers ran alongside the ferries in a joint military exercise to cross the bay. Each form of transportation symbolized a new future for the United States and its global standing. The Pacific Fleet was composed of fifty-two vessels that held over fifty thousand crewmembers under the command of Admiral Hugh Rodman (who was housed on the USS *New Mexico*). Much as the convoy had garnered crowds on its journey, the fleet had been followed up the coast by over one million onlookers. The festivities of early September were a convergence, celebrating the "land armada" and the "sea armada." Oakland's truck dealers had sent out a notice about the importance of this gathering: "The people of the United States look to California to do big things in a big way, and this is an opportunity we cannot afford to miss."[149]

Once the convoy's vehicles had disembarked from the ferry, they lined up in formation on the Embarcadero (the grand boulevard that runs along the city's waterfront) and at 11 a.m. they set off toward Market Street with their military and civilian motor escorts. Moving through the city streets of San Francisco brought a different type of dissonance. These streets were certainly not intended for such mammoth machines, and the normal cacophony of city activity was directed on the important moment. Although the heavy vehicles had had many difficult passages through the unpaved western roads, most of them were better suited for those conditions than an urban setting. Despite such limitations, parading through the middle of San Francisco, the convoy was treated to its largest celebration yet. Residents waved from open downtown building windows, cheering as the parade moved over an eight-mile track, passing the Civic Center, moving along Presidio Avenue and through Japantown to Fillmore and, finally, ending at Lincoln Park. There, at the Palace of the Legion of Honor overlooking the Pacific Ocean, a stage had been placed for the closing ceremonies. And, of course, with boosters fully cognizant of narrative, there was also a milestone to match the Zero Milestone placed months prior near the White House in Washington, DC. The marker in San Francisco, though, read 3,251 miles.[150]

At the ceremony, McClure presented the mayor and governor with the laurel wreath he had shepherded across the continent. The LHA had produced commemorative medals for each of the members of the convoy (gold for the organizers and bronze for the troops). Never one to miss an opportunity, in his speech Dr. Johnson urged for federal funds to support construction of the LH and to establish a national system of roadways. After all, this entire operation had not been without purpose: better than any written declaration, the slow, plodding convoy had demonstrated in real time the inefficiency of the nation's so-called system of roads. A national focus of attention and funds had been dramatized—if Americans wished to accept the challenge. And, finally, the task concluded: the trucks were sent into service and troops to posts in the area.[151]

Of the convoy's participants, the *Oakland Tribune*, wrote, "Theirs was a man's job and they did it well. . . . They have proved once again that what Americans decide to do, they do, despite all odds of nature and mankind." In its coverage of the event, the *San Jose Mercury News* also plugged a specific vehicle:

> The test is a momentous trial of the motor vehicle as an agency in peace time, as well as in war. The tour marks government recognition of the importance of good highways and dependable motor vehicles. . . .
>
> The cross-country trip of this first motor transport convoy was in this way, as historic an event as the first trans-Atlantic flight. . . .
>
> In a test consuming several weeks and necessitating driving every day, great care was naturally exercised in the selection of cars to carry observers and passengers to the end of the greatest degree of comfort consistent with car endurance and flexibility. . . . The choice of the new Overland model as the carrier of officials and newspapermen was a tribute to the comfort quality which the Willys-Overland company claims for its new type.[152]

In truth, though, not every part of the convoy had completed this most difficult stretch of the trip. Over the final few weeks, the haul that brought it to San Francisco had presented some of the most challenging terrain of the entire trip.

"THERE IS plenty of mud," McClure cabled ahead gloomily as the truck train crossed from Nebraska into Wyoming at Pine Bluffs on August 8. Given the terrain that separated the convoy from its finish line, mud was the least of the worries facing the drivers.[153]

Crossing the Red Desert in the center of Wyoming's sheep rangeland, McClure described the scene as "most desolate and

Bridges failed in a variety of fashion and required on
the fly engineering solutions.

[monotonous] character" defined by "dry air wind and dust
hardship continuous." Bridges requiring replacement and a water tanker flipping onto its side all slowed their progress, and the
convoy covered a mere fifty-eight miles in eleven hours. By this
point, engineers on the crew had become adept at applying their
battlefield skills to quickly and effectively securing or replacing
bridges as needed, so the primary cause of the slow pace was
certainly the terrain.[154]

Beyond the western branch of the Missouri River, desolate
plains become craggy and eventually mountainous. Wind storms
stirred with gusts of fifty miles per hour and made it nearly impossible to remain on their path, which in Wyoming was at best
packed gravel and often still little more than a trail. McClure described a scene of "utter desolation strewn with bones of animals.
The intensely dry air absence of trees and green vegetation and
parched appearances of landscape exert depressing influence on

personnel." In summary, he referred to the Green River area of Wyoming as "altogether most tedious day of expedition."[155]

More than at any other point in the trip, the convoy here was crossing ground previously untrammeled by almost any form of human transport. During one day of this passage, they reinforced four bridges, rebuilt a timber culvert, and detoured around four others. On another day, they stopped nine times to fix bridges and culverts and didn't dare to try one other bridge. Instead, one by one the trucks either drove slowly or were lowered with winches down the vertical cliffs near the creek and were hauled up on the other side. And these problems of terrain were only exacerbated by the political topography of Utah. Debate over the location of the LH had preceded the convoy, and a fluid plan had not even been entirely settled as the truck train approached.[156]

Even where the LH was plotted across the state, it presented confounding topographical quandaries that were linked to economic motivations of the planners. Three options for its route vied to be deemed the most practical—none being particularly stellar. One option was to follow the original line of the Union Pacific Railroad from Ogden, moving north to go around the Great Salt Lake. This meant it bypassed Salt Lake City and proceeded on a northerly route through Nevada to intersect more directly with San Francisco. Another option was to pass the Great Salt Lake to the south, but still orient travelers toward San Francisco. Most motorists leaving Salt Lake City, though, were destined for Los Angeles, toward the south. A final option, which was the winner, dropped even further south than the second and followed the old Pony Express route. When that route was selected in 1913, the logic was that it moved south around the worst of the desert (keeping travelers within twenty miles of water throughout), ran through central Nevada (instead of the north); travelers could arrive in Ely, Nevada, and choose between following the LH to San Francisco or the former Midland Trail to

Los Angeles. Indeed, the traveling was made even more difficult in 1919 by the fact that the accepted route was still being debated and remained incomplete.[157]

Even after they finalized a route for the LH, there was another obstacle: learning how to build a road on the region's remarkably and uniquely problematic ground. An odd combination of sand and rock, the Utah ground fluctuated significantly with any rainfall and had proven entirely unreliable to support most types of construction used in other regions. Natural obstacles such as the Stansbury Range and the salt desert created impediments to any planned route. Even the dry stretches were composed of packed dust that was subject to blowouts when wind literally tore out sections of the passage.[158]

This stretch of Utah, though, had begun wrestling with the implications of the modern world years prior.

Big Bill Rishel, a popular observer of the West and early cycling enthusiast often referred to as the "desert bedouin," was one of the main players in Utah's road-location drama. Faced with the challenge of the Stansbury Range, an earlier group had investigated a path through the Tooele Valley to Timpie Point in 1913. It became known as Johnson's Pass after one of the first settlers of the town of Clover, which was nearby. In 1915, a group toured the pass and ended up snowed in at Orr's Ranch in Skull Valley. Still, the route's proponents argued to the governor that it should be chosen, and Henry Joy even drove his Packard through it to prove that it was not only large enough to fit horse traffic but passable for cars too. He worked with Henry Ostermann to get the route approved and to widen Johnson's Pass, which he argued would shorten the trip by fifty miles. But these improvements would require capital investment, and that's how Big Bill became involved.[159]

In this debate over the LH routes through Utah, Rishel squared off against Goodyear Tire & Rubber Company founder

In the Salt Lake City reception, even the floral float was made to resemble one of the convoy vehicles.

Frank Seiberling. Once the automobile arrived, Goodyear, of course, became a massive company, and by 1919 Seiberling employed twenty-five thousand people and had sales of over $1 million per day. He promised $75,000 from Goodyear to pay for the Johnson's Pass improvement and then went to visit the site in 1917. Repeatedly his group got bogged down and had to dig themselves out from the salty mud. The trip was so terrible that Ostermann and others expected Seiberling to entirely pull out of the project. Instead, he proclaimed that $75,000 was insufficient and agreed to pay $100,000. In agreement with the state of Utah, Goodyear funded the job in 1918, and furious work continued through July 1919, in hopes of preparing Johnson's Pass to host the convoy's passage. From Detroit, communication was sent out to coordinate with the convoy to create a grand opening event for the Pass. The main contact was the secretary of the Utah State Automobile Association, Big Bill Rishel.[160]

While a coordinated event was eventually dashed due to challenges of timing, influence was still exerted to steer the convoy through Utah. Rishel believed the Johnson's Pass route would permanently undermine development of the LH. Until the bitter end, he fought to direct the truck train to one of the alternative routes. By the time the convoy set out, Rishel supposedly vowed, "If the army tries to cross our [highway], I will stand guard with a shotgun and defy them. By all means, let those big trucks over the 'Seiberling Section,' and there won't be any Lincoln Highway anymore." In the coming days, the convoy followed the "Goodyear Cutoff," as it was now called. And that, actually, proved to be the best way for Rishel to accomplish his ultimate goal of directing the LH's location.[161]

SOLIDIFYING ANY permanent path through the shifting terrain of Utah required importing enormous tonnages of rock and gravel, which was a costly addition to any road construction. The passages traveled by the convoy in 1919 were barely as wide as the trucks themselves and wound between red cliffs on each side. When the passage wound higher along the ridges, the soldiers often had to reinforce the crumbling roadway in hopes of allowing the heavier vehicles to hold the road and not to slide off down the hillside. Beyond Rishel, some local proponents wanted the truck train to be the first user of the new roadway, but the timing presented a daunting challenge. As the convoy closed in, only seven miles of the seventeen in the cutoff had been graveled.

Heading for Salt Lake City from Ogden, the convoy traveled road that Utah had not been able to afford improving at all. It was particularly in these stretches that passage was only possible through the expert usage of the Militor. First, the flexible vehicle towed the machine shop (which had blown a piston earlier) when the trailer broke through the crusted road surface and into the mud. Then, as the Militor driver tried to pull the machine shop out, it also broke through and sunk over four feet into the mud below. Truly challenging times when the wrecker is stuck as well as other vehicles! The three-hour process to free the Militor broke steel cables and sheared off a pulley. Changing their approach, the soldiers used shovels to remove the mud and installed planks that allowed the wrecker to drive out of the crater. Once the Militor was freed, they used the same route for it to then tow out the machine shop. Clearly, the Militor had solidified itself among the men as the most essential and effective device in the convoy, particularly in the unsettled roads of the West.

After a tremendous celebration in Salt Lake City, the truck train left town at 6:30 a.m. on Wednesday, August 20. Following the travails of the previous days, three trucks remained behind for repairs and, most problematically, the heroic Militor stayed with them in order to ensure their travel once they again got on the road. It was a judgment call, and, in one of the first examples of dissent related to the convoy, not everyone agreed on the decision. The Militor was simply needed in too many different locations. This episode presents one of the clearest fissures of the convoy, this time between McClure, public relations concerns, Elwell Jackson, and the technical considerations of carrying out the cross-country excursion. Ironically, though, once the repaired vehicles continued on the trek, it was the Militor that repeatedly broke down, forcing roadside repairs.[162]

There had been a dry spell, no rain for some eighteen weeks, and the fine grit of alkali dust and sand swirled throughout all the vehicles on the road beyond Salt Lake City. The stage was set for a grueling day. First, in the air, grit infiltrated the vehicles' air systems, bearings, valves, brakes, and clutches, impacting their effectiveness. Then the trucks sank into the dust with no traction, until they were buried in loose sand up to their chassis—and often got stuck. And, of course, the Militor had been left behind with the vehicles under repair in Salt Lake City. With the Militor unavailable to this segment of the train, each stuck vehicle had to be jacked up and dug out by hand. Once a vehicle was freed, men stripped sagebrush from the desert and laid it in the wheel ruts to provide something that the tires might bite into. After seventeen hours, the trucks pulled into their next stop at Orr's Ranch; however, four trucks had been temporarily left behind in the desert. Meanwhile, observing their diminishing supplies of gasoline and water, McClure had grown particularly worried about the absence of the Militor, which had not caught up, and his frustration would only increase.

Elwell Jackson had left Salt Lake City about four hours behind the main body of the truck train and had come upon the Militor disabled by the side of the road. The Militor, overseen by Sergeant Theodore Wood, had run into its own mechanical difficulties. For some time, Wood had managed to improvise a solution to the wrecker's ongoing problem with a fan belt, but it had finally given out. Wood was frustrated that his repeated requests for a new belt had gone unheeded. Jackson went on ahead with the other trucks that had been repaired and assumed Wood would fix the Militor and catch up. But Wood experienced even more mechanical problems and limped along until he was finally able to get the wrecker to a garage in Tooele, after which he hoped to catch up with the convoy at Orr's Ranch. But, leaving Tooele, the Militor failed again, and Wood and his crew spent the night under the Utah sky.[163]

The next morning, as the rest of the convoy departed Orr's Ranch, Wood returned to Tooele for repairs to the Militor. McClure backtracked to Toole to check in on the Militor and in his frustration advised Wood to take the wrecker back to Salt Lake City for an overhaul and then to ship it by rail to Eureka, Nevada, where he might rejoin the convoy. Wood, who had by then almost completed the repairs, argued that he'd rather set out shortly and rejoin the truck train. McClure insisted that "that this was impossible." Following his orders, Wood returned with the Militor to Salt Lake City. He spent five days re-conditioning the Militor before leaving with it by train. The convoy's greatest hero machine was now out of action. This most valuable vehicle, which had rescued every other vehicle in the convoy at least once, had been benched two states shy of the goal. What would it mean for the rest?[164]

When Jackson joined with the rest of the convoy at Orr's Ranch, he learned of the fate of the Militor—and Wood—and was incensed. In his next cable to Washington, Jackson called McClure's decision "entirely unwarranted" and he later clarified that in his judgment, the Militor was "unquestionably the most valuable vehicle in the entire convoy." Obviously, the terrain and mechanics had impacted the military order that had held thus far. Later in the day, Jackson rode ahead of the convoy from Orr's Ranch to study the road conditions that lay beyond. He was mystified at how the convoy would manage passage without the invaluable Militor. Indeed, he was proven pretty correct in his assessment. After leaving Salt Lake City, the convoy moved at the slowest rate of their entire trek thus far and did not reach their makeshift evening camp site until 10:30 p.m. And the following day was even slower.[165]

At 6:15 a.m on Thursday, August 21, the truck train set out from the ranch for the Goodyear Cutoff to head southwest. This drive carried them through the windswept salt flats and alkali barrens. Incomplete road construction forced the vehicles to detour directly into the featureless salt. The slim crust gave way under each vehicle, sinking them into a thick silt that ensnared their wheels without release. Even the tracks on the caterpillar tractor were useless. The age of road travel had not yet arrived on these salt plains, and human power was the only answer. Passage was only possible by tying a rope to each vehicle, one by one, and having a row of fifty to one hundred soldiers pull each vehicle forward. Jackson observed: "Practically every vehicle was mired and rescue work required almost superhuman efforts of entire personnel from 2 p.m. until after midnight." Their average speed on this day was just two miles per hour.[166]

McClure messaged ahead to Gold Hill to inform Washington of their dire situation. "Mere existence was chief concern," he reported. Meanwhile, each vehicle struggled through a second day mired in the flats. The entire group was also running dangerously low on water and fuel. McClure placed the single water truck under guard, and soldiers were rationed a single cup full overnight. Assistance arrived from a local road construction crew whose superintendent, Walter Paul, used horse teams to haul two tanks of water twelve miles from Gold Hill to Black Point. McClure reported that the soldiers were "utterly exhausted by tremendous effort." Their morale, he wrote, had been crushed by the Salt Lake Desert and its "utter desolation and isolation/the excessive heat and glare/absence of live vegetation and water/silence/deceptive mirages and apparent lack of means of exit."[167]

The struggle through the salt flats ended when the final vehicle arrived in Gold Hill around 2 p.m. While their schedule had been totally missed, McClure allowed the men much needed rest before departing for Nevada the following morning. Behind them, the Goodyear Cutoff was left in tatters. Rishel described the roadway in this fashion: "There were ruts hub deep and holes large enough to bury an ordinary touring car. The new grade looked like a terrain shelled by modern bombers. There were almost enough new planks buried in the mud to build a bridge across the mud flats." At least temporarily, the essential link westward for the LH was left impassable—destroyed by the convoy that was intended to publicize it.[168]

*   *   *

STAGGERING OUT of the challenging mountains, the convoy hit much flatter terrain across Nevada and reached Fallon on August 29. Though worn out, the crew was relieved and gaining hope that the end was nigh—even if their progress remained painfully slow by modern standards. After leaving Fallon, they required eleven hours to drive twelve miles. Any effort at scheduling had now become irrelevant: not only was the convoy a full five days behind its hoped-for pace, but the roads of Nevada were largely nonexistent. In this remote spot, the LH was a line drawn on a map; however, at least half of the proposed road system hadn't even yet been surveyed. Jackson recorded, "All heavy vehicles, including Cadillacs had to be pulled and pushed through by convinced efforts and men, over wheel paths made of sage brush."[169]

Referred to as "The Loneliest Road in America," the 250-mile stretch of the LH between Ely and Fallon in 1919 held only two settlements, Eureka and Austin, each with populations of less than one thousand. Leaving Ely at 6:30 a.m, the truck train inched across the sand flats of the Long Valley in ruts in the dust that measured as deep as eighteen inches and wind gusts of twenty miles per hour. Slowed by the winding roads up the Diamond Mountains, they stopped for the night just short of Eureka, and the soldiers simply slept along the road. The following morning, they stopped only for a half hour in Eureka—long enough for Jackson to note that the Militor had not yet arrived to meet them—and to apologize that they had missed the planned festivities the night before. In his explanation, McClure noted that the convoy was "on the last end of a decidedly hard journey."[170]

With mechanical failures mounting, the truck train stopped late in the afternoon, shy of its intended goal. "Remarkable that equipment remains serviceable," McClure reported, "with abuse given by these deplorable roads." Able to make a few adjustments

Crashes occurred frequently and involved on the fly
repair and extrication.

and repairs in the small settlement of Willow Spring, they arrived
in Austin, Nevada, early in the afternoon the following day. A
prototypical western town from the days of silver mining, Austin
welcomed the convoy by converting four of the jail's empty cells
into showers for the soldiers. Next, the train proceeded through
the sandy stretch known as the Fallon Sink and straggled into
Carson City. Other than Utah's salt flats, McClure reported that
this stretch through Nevada had been "the most unfavorable
combination of road conditions yet experienced . . . the most trying day of the trip."[171]

Fortunately, a great celebration in Carson City awaited the
dejected, exhausted soldiers. Linen-covered tables and bands
filled streets decked out with strands of electric lights. Treated to
a chicken dinner, the men then spread their cots on the lawn of
the capitol building. The local *Daily Appeal* reported, "Carson

extends the hand of welcome to the visiting members of the government caravan and trusts that in their two days' stay they will get a partial rest at least from their labors. They have made one of the greatest trips on record and it has been a test of their nerve, their endurance, and their patience. But it has proved the American spirit and what it is capable of." During their trek, the soldiers had been celebrated by over three million Americans in 350 different communities, but Carson City stood out. The rest day to follow helped; however, the travelers must have also drawn great solace from knowing that they were past the most challenging and unknown stretch of their journey.[172]

There were still difficulties to follow, though. As they made the ascent of the Sierra Nevada at King's Grade, the heat proved nearly unbearable. Temperatures were recorded as high as 110 degrees. "Usual mountain and desert trails," Jackson reported as he barely hid his fatigue; McClure's tone was just as jaded: "Usual deep dust sand chuck holes and ruts." Even though their process was painfully slow, the crew had planned for this ascent. To manage the steep grade, one man on each vehicle was poised and ready to block the wheels each time the convoy halted. Jackson described the ascent in this fashion: "Reached altitude of 7,630 feet at summit, over narrow, winding road of sand and broken stone, cut out of, and, in places, built up on a mountain side. Total climb 14 miles made in 6 hours, slow progress being necessary to prevent accident. Grades 8% to 14%. Crossing Sierras without accident may be considered noteworthy achievement for heavy vehicles."[173]

These were the steepest climbs of the journey, and the Militor was still unavailable to help. For the most precarious mountain roads, McClure sent the tractor up ahead to stand ready to assist other vehicles. He also set up a checkpoint at the foot of major inclines, where each vehicle was carefully inspected. Then he sent them up heaviest to lightest, with one hundred yards between them. They traveled slowly enough that a soldier could walk alongside, with wheel blocks at the ready to stabilize any vehicle that started to slide.[174]

Now, the end was in sight. From Placerville to Sacramento, Jackson wrote in his report with a marked change in tone: "Entire route down grade over bitumen surfaced concrete roads lined with palm trees, through peach, almond, orange, and olive ranches and vineyards. Populace showered convoy with fruits. . . . Fair and warm. Perfect roads." With that last phrase, he clearly meant to emphasize all the imperfect roads that they had also passed. Crossing into California on September 1, they had been on the road for fifty-seven days.[175]

IN SOME of the best coordinated planning of the venture, the California State Fair had begun in Sacramento on August 30 and was easily refocused to serve as the largest celebration since the convoy had left the Ellipse.

"Please show us a good road," McClure told those who came ahead to meet the convoy in the mountains. Much of the roads that they had passed, he explained, "belong to the Stone Age. It is impossible for worse roads to exist." The communities of the far West had progressed much more quickly, indeed. Ahead, he wrote later, they found "a revelation in good roads" and the truck train arrived in Sacramento at 2 p.m. on September 1.[176]

In Sacramento, the Willys-Overland Company hosted a grand four-hundred-plate dinner of razor-clam chowder, Sacramento River salmon, country-fried chicken, and "Overland Ice Cream," which the crew enjoyed with dignitaries. In his comments at the event, McClure sounded much like a man near the end of his mission: "I was sent to bring the train through, and we are nearly done. . . .We have shown that it is possible to get through. I believe good roads will follow us." In the dinner program, John Willys, Willys-Overland president and LHA director, compared the participants of the convoy to the "immortal 'Forty-Niners,'" writing, "Their blood is the blood of the western country: strong—virile—self-reliant." Like those early western settlers, he said, the convoy had blazed new trails "of Commerce, Highways, Mechanical Achievement, and the Protection of the Flag." The evening concluded back on the fairground with a dance in front of the General Exhibits Building.[177]

Resisting the offer to remain at the fair for another day, the convoy left the following morning and stayed over next in Stockton, California, and then, finally, in Oakland. In preparation for their final stop, the soldiers were outfitted with clean new uniforms, and McClure gleefully and with relief described the final seventy-six miles of the LH as "unexcelled."[178]

Officials in Oakland had been following the convoy's progress for two months while they planned their own festivities. Alameda County's Automobile Trade Association took the lead, and maps were produced to show the public the route that the convoy would follow through their region. Every vehicle owner in the East Bay was encouraged to form a giant avenue of automobiles past the Chevrolet plant on the outskirts of town—what one organizer referred to as a "continuous ovation" of vehicles and drivers. Indeed, as planned, thousands of motorists met the truck train and a parade of bands and floats met them ten miles east of Oakland. In Jackson's words, they were "escorted through Court of Honor and flag festooned streets, while all whistles around Bay were blowing. Elaborate electrical and fireworks display. Dinner, Hotel Oakland. Dance, Municipal Auditorium." Described by McClure as "tumultuous," the reception included a rocket shot into the sky that rained Allied flags down on those gathered along the streets. The officials and soldiers in the convoy spent their final night with a grand celebration before heading to Oakland Harbor in the morning.[179]

*   *   *

AFTER PASSING the Pacific Fleet while being ferried across the Bay on the morning of September 6, the "land armada," as it had come to be called relative to the great naval gathering in San Francisco Bay, moved through the streets of downtown and arrived at the entrance to Lincoln Park, where the new marker had been placed to commemorate the completion of their journey. A pair of army planes circled overhead and showered the scene with flowers, while the battleship *Arkansas* sailed just off shore. It was a fitting symbol of the remarkable military standing with which the United States had emerged from World War I.

Just before 2 p.m., McClure sent his final message: "CONVOY ARRIVED FINAL OBJECTIVE SAN FRANCISCO TEN MORNING ALL EQUIPMENT ROLLING NO MECHANICAL DIFFICULTIES OR CASUALTIES DETAILS REPORT LATER."[180]

The reality, of course, was a bit more arduous to detail. Of the original eighty-one vehicles, six did not finish. Apart from the Militor, they'd lost one truck on the mountainside in Pennsylvania, another had broken down beyond repair on the Goodyear Cutoff, and three of the four kitchen trailers had been practically demolished by bad roads. The Militor arrived in San Francisco five days behind the rest of the convoy. Sergeant Wood had finished overhauling it and the railroad had taken two weeks to haul it from Salt Lake City. The trucks were sent for service to the Army's western division, the men to posts around San Francisco.

Captain William C. Greany had served as McClure's adjutant and statistical officer throughout the expedition, and he was tasked with creating a final report on the convoy. In his report, he described the soldiers' living conditions in the convoy as much like "those generally experienced in the advance zone of battle operations, but the tour of continuous duty was of a longer duration than is usual for such service." According to historian Kevin Cook,

[Greany] said that the convoy's personnel got an average of only five and a half hours of sleep per night. Greany figured that "1,778 miles or 54.7 per cent of the mileage was made over dirt roads, wheel paths, mountain trails, desert sands, and alkali flats." He added that 500 hundred miles was practically impassable and required "combined efforts of the most extraordinary character." He counted 230 road accidents, which included "instances of road failure, and vehicles sinking in quicksand or mud, running off the road or over embankments, overturning, or other mishaps."[181]

Certainly, the mission's completion merited great recognition.

Johnson and others spoke eloquently at the festivities, about the convoy and of the larger need for a national system of roads. Indeed, the journey was a monument to the commitment to a transition in American life, particularly as it related to energy and transportation. The *Oakland Tribune* speculated "that if within two years the tour were to be duplicated over the same route, the Lincoln Highway, it could be made in one-half the time. Our feeling is that boulevards will exist in stretches now noted for their horrible condition. A concluding impression is that the land fleet tour has illustrated, very forcibly, the necessity for passing the Townsend Bill." It seems clear that, indeed, the moment in San Francisco was meant as both a conclusion and a beginning.[182]

THE CLOSE of the convoy continued to stir the efforts of the boosters who had been behind it. For them, its close was also a symbolic start to a new era. Henry Joy wrote, "It is given to very few mortals to see their dreams come true. Especially is this true when for realization those dreams require the awakening of a whole people to a new order of things. . . . [Yet] the Lincoln Highway, in reality nothing but a dream in 1913, and by many thought to be a very wild and impossible one, is coming true."[183]

When the convoy disbanded, Henry Ostermann drove back to Detroit with Captain McMahon, with whom he had hatched the entire plan for the convoy. Soon after his return to Detroit, he married a second time and—of course—took his bride on a drive west to celebrate. He was tragically killed in an auto accident outside of Tama, Iowa. The Lincoln Highway cause was carried forward by Gael Hoag, its new field secretary. Most importantly, to complete the important leg through Utah, five years of debate ensued before Rishel's influence guided the completion of the Wendover Route instead of the Goodyear Cutoff. The route was completed in June 1925, and legions of Americans flocked to buy new autos. While the new drivers appreciated the vehicle's daily convenience, many also aspired to take their wheels to the open roads of the West in due course.[184]

Also, it can't be overlooked that the convoy had inextricably linked the highest echelons of the military with the push for federal support for road building. When he was sworn in as President in March 1921, Warren G. Harding was the first US president to travel not in a carriage but an automobile—a Packard Twin Six, to be specific. Driving and roads clearly had become a symbol of a modernizing nation. National progress, it seemed, came on four wheels and was powered by cheap gasoline.

At the ground level, though, such infrastructural change remained a bit daunting, and over the next decade the LH limped forward as a large-scale initiative. By 1927, Hoag wrote that the LH was hopelessly "scrambled." Seiberling lost his fortune by the end of the 1920s as Goodyear massively restructured. The influence of Ostermann and others waned, and by late 1927 the LHA dissolved and closed its offices in the GM building in Detroit. Their moment of boosterism stands as a true tipping point in the American energy transition of this era, but the fight would pass on to others.[185]

Henry Joy died in 1936, and in the intervening years he had become only more boisterous but irrelevant. He used the new medium of radio to seek out an audience for his criticism of many government activities and even established his own radio station to broadcast from Joy Ranch. But few really listened or cared. He was particularly public in his stance against Prohibition, which drove him from the Republican Party. A few years after his death, his wife erected a roadside monument in his honor in the emptiness of Wyoming. Its inscription read, "That there should be a Lincoln Highway across the country is the important thing."[186]

Most of the financial backers of the convoy suffered mightily during the Great Depression of the 1930s. Of the participants in the convoy, the biggest immediate winner may have been the Packard automobile. The company was the only luxury car maker to survive the economic downturn, and then during World War II, Packard shifted into manufacturing airplane engines. In 1956 the company merged with Studebaker and the final Packard was built in South Bend, Indiana, in 1958. Joy's wife, Helen Hall Joy, died in 1958 as well. An interesting footnote: as she became the grand dame of Detroit motoring, she always drove a 1914 electric vehicle—a navy blue brougham—that Henry had bought for her.[187]

PHOTOS PUBLISHED nationally displayed to American consumers the great celebration in San Francisco at the convoy's close and the remarkable cast of characters who had made it and who had made it possible—except for at least one. Eisenhower missed the celebration. When he had seen Mamie and her father in North Platte, Nebraska, he had determined to put in for a month's leave so that he could spend time with her.

Despite his two months on the road, once the convoy reached San Francisco, Eisenhower set out to reach Mamie and her family in their winter home of San Antonio, Texas. The rains were so severe during his trek that he had to stop and wait for a week in Oklahoma City, until the roads were dry enough for safe passage. Of this period, he later recalled, "There were moments when I thought neither the automobile, the bus, nor the truck had any future whatever."[188]

In the coming months, though, the experience of the convoy settled into the subconscious of young Eisenhower and the entire country. Its true impact would come years later.

During the 1952 presidential campaign, President Dwight D. Eisenhower often drove the early electric car that had belonged to his wife's family. The 1914 Rauch & Lang, with swivel seat, removable brass clock, and crystal bud vase, had been purchased new by his father-in-law, John Sheldon Doud, a prominent Denver businessman, for his wife, Elivera.

CHAPTER FIVE:

Auto Nation

I t was very likely a challenge to be Eisenhower's pal. Once he made his choices, the events that transpired often seemed almost like a fairytale of happenstance. Not that he was an un-committed friend to everyone in his orbit—quite the contrary. But once he steered onto a life path, associations were simply struck with the greatest military and political leaders of the century, and challenges and opportunities were thrust upon him on which countless human lives hinged. What began as simple personal choices ultimately resulted in large-scale change, with Eisenhower creating what Henry Luce and others declared the "American Century."

Placed in the context of the twentieth century, the 1919 convoy was a significant historic achievement for everyone who participated. Even if one of the participating soldiers had not been a future US president, the convoy would still have been noted in history. But that is not where the story ends. And the future president was not simply a bystander in the century that unfolded after the trek had been completed.

Today, we can look back and see that Ike, after his participation in the convoy and subsequent choices as a national leader, was a catalyst. And, if we attempt to connect the dots that led him to choose as he did to re-take Europe in World War II and to bring forth a new culture of roads at home, his time on the convoy appears less as a "lark" and more as an undeniably formative event.

After the convoy, the nation steered itself into the passing lane on the figurative highway of energy consumption and depressed its accelerator without fear, question, or trepidation. At the center of this transformation is Eisenhower, the young man from Kansas. The transition in American travel in 1920, just after the convoy, will prove to be the defining point for how future generations of humanity will recall American life. It is when we chose to hit the road.

Eisenhower associated with the great U.S. military leaders of the time, including General MacArthur.

Ike in World War II camp.

Two decades after the plodding progress of the convoy, Ike was thrust into a rapid-fire moment that would define his remarkable life. No other human in history has lived a moment like what Ike experienced between 1941 and 1945.

In the pivotal window of time after 1936 when Adolph Hitler and the Axis powers blew through Europe and sought to dominate the world, it is no overstatement to say that Eisenhower devised the method by which the Allies would stop them in their tracks. Simply put, without him and his plan—first formulated while sitting alone at his typewriter in Washington in December 1941—the outcome for the free world and for the entire century would have been very different.

In the process of this remarkable achievement, Eisenhower's long-term perspectives were solidified. He discovered global models that took the inspiration he had found in the convoy to new levels. In particular, his time on the roads of Germany during and after World War II left him with a clear vision of what could happen in the United States if centralized planning and strategy seized on the innovation of auto travel. This vision had begun to take shape in 1919 with those hours spent in the Packard's bucket seat watching inadequate roads repeatedly challenge the convoy's vehicles. Truly, by midcentury, he had become a unique vessel for moving the germ of the vision that had spurred the convoy's boosters into a full redesign of the American economy and landscape.

DURING THE decade that followed the convoy, Eisenhower chose to remain in the military. He completed Command and General Staff School (CGSS) at Leavenworth, Kansas, and then accepted a remarkable string of opportunities to work with some of the greatest military leaders of American history. Although he didn't pursue leadership, it seems to have pursued him. These leaders observed his potential and commitment and did not shy away from assigning him increasingly significant tasks.

For fourteen of his thirty-seven years in the military, Ike worked directly under George C. Marshall and Douglas MacArthur, each of whom holds a place in the pantheon of the modern era's greatest generals. And, still, so many of the pivotal events of his life came about through happenstance. For instance, following CGSS (where he graduated first in his class), Eisenhower was assigned to the War Department as a staff officer to General John J. Pershing, who put him to work preparing a history of the American army in France. So pleased was he with Ike's output that Pershing next sent him to the Army War College

Consulting in leadership decisions defined much of
Eisenhower's war-time service.

for a year and then to Paris to expand the history and study on
the actual ground. He and Mamie found an apartment at 68 Quai
d'Auteuil, near Pont Mirabeau, on the Left Bank of the Seine.
Returning to Washington in November 1929, Eisenhower served
as an aide to the new chief of staff, General Douglas MacArthur.
In a fateful observation, MacArthur wrote of Eisenhower in the
early 1930s, "This is best officer in the Army. When the next
war comes, he should go right to the top." Thus, Ike—largely by
chance—gained a unique expertise regarding France, Europe,
and military history in general.[189]

Although Eisenhower hoped to gain experience directly with
troops during the 1930s, MacArthur would not allow it. When
his appointment as chief of staff ended in 1935, MacArthur went
to serve as a military advisor to the new leader of the Philippines
in Manila and demanded that Ike join him as his assistant, a post

at which they remained until 1939. The experience, though not what Eisenhower had hoped for, again provided him with unique knowledge of an entirely different military theater. In short order, his breadth of knowledge had become unparalleled, and it fed the military genius that he had perfected over his years of training.[190]

When World War II began in September 1939, Eisenhower wrote to his friend Leonard Gerow concerning modern warfare, "It seems obvious that neither side desires to undertake attacks against heavily fortified lines. If fortification, with modern weapons, has given to the defensive form of combat such a terrific advantage over the offensive, we've swung back to the late middle ages, when any army in a fortified camp was perfectly safe from molestation. What is the answer?" Indeed, it would largely be his own task to answer the question he had posed. Having worked in various staff and leadership roles during the frenzy to prepare American forces for war, Eisenhower possessed clear, diverse expertise; however, he also appeared poised without fear to use weapons and force when it was needed.[191]

On Sunday, December 15, 1941, Eisenhower responded to a summons by General George C. Marshall and arrived in Washington, DC, for tasks unstated. Now fifty-five years of age and feeling mired against the advancement he desired in his career, Colonel Dwight Eisenhower arrived at Union Station in the morning. He did not know why the chief of staff of the army, a close colleague in leadership, had summoned him to Washington. The country had carried out an intensive military buildup over the two years since World War II had been declared, and, now, after the Japanese attack at Pearl Harbor and President Franklin Roosevelt's declaration of US involvement in the war only a week earlier, the sense of foreboding was palpable.

Ike arrived quietly by train, without fanfare—for very likely the last time in his life—and proceeded directly to the War Department (the Pentagon was under construction). After a general introduction and description of the losses at Pearl Harbor, Marshall leaned over his desk to look directly at Eisenhower and asked, "What should be our general line of action?" The young soldier's life would never again be the same. "Give me a few hours," he stammered in response. Marshall agreed. Ike proceeded to the desk that had been temporarily assigned to him in the War Plans Division of the General Staff Office, placed a sheet of yellow paper into his typewriter, and with his one-finger style typed, "Steps to Be Taken."[192]

In response to Marshall's summons, Eisenhower only expected to be asked about what he had been working on most recently: force strength in the Pacific, primarily in the Philippines. Instead, at his lonely typewriter, he followed Marshall's orders and produced a general description of what the US military should do next on the global scale.

Later that day, when Ike presented his thoughts to Marshall, the general responded simply, "I agree with you." By Ike's recollection, Marshall then leaned forward and said, "Eisenhower, the Department is filled with able men who analyze their problems well but feel compelled to bring them to me for final solution. I must have assistants who will solve their own problems and tell me later what they have done."[193]

In this remarkable moment of historic convergence, greatness was not forced on Ike in a specific moment on the battlefield; instead, when opportunities were offered, he excelled, and the magnitude of the assignments catapulted his stature to levels that literally knew no borders. By the end of 1941, Ike was promoted to major general and Marshall took him along as his chief assistant to the first wartime conference with the British, assigning him to prepare the basic American position on organization and strategy for a global war. In early 1942, Marshall made Eisenhower the head of the War Plans Division, which was then renamed the Operations Division, with Ike as its commander and more than a

Ike in World War II walking in camp among soldiers.

hundred officers working for him. By 1943, Eisenhower had grown convinced that war in Europe was essential, and he drew up the plan known as "Roundup," later evolving into "Overlord." Marshall received approval for the plan, first from Roosevelt and next from Prime Minister Winston Churchill and Britain. To carry it out, the Allied nations established a joint office of the European Theater of Operations (ETO), based in London with Eisenhower as its commander. He arrived there in June 1943.[194]

Of course, the successful efforts of the ETO, the inspiring D-Day invasion, and the eventual liberation of Europe in 1945 christened Ike the strategic genius of the forces that turned back the Axis threat. Less remembered, though, is that Eisenhower also presided over the aftermath of the fighting and the recovery of Europe.[195]

While his convictions remained on the importance of family, the value of new weapons systems, and the modern use of roads, the midwestern lad from Kansas reached the late 1940s as one of the most recognizable figures on earth. Britain declared him an honorary citizen and bestowed on him its highest honor, the Order of Merit; most European nations offered similar honors. On June 16, 1945, Eisenhower returned to Washington for the first time in nearly two years and met with President Truman, addressed a joint meeting of Congress, and proceeded to New York, where at city hall more than four million—the largest crowd in the city's history—greeted him. A few key choices—ranging from joining the military, embarking on the convoy, and committing potential leadership possibilities—initiated Eisenhower's c A far cry from his quiet arrival to Union Station a fe

FROM TODAY'S perspective, Ike's next steps in the 1940s, politics was not a certainty for party affiliation remained a mystery to m spent the months following VE Day p authority administration of Germany, familiarity with the nation's modern the Autobahn. While Ike was preoc cover and stabilize Europe, postwa returning GIs and a profound na ization and economic expansion war of ideology with the USSF the American middle-class lif other nations that the US me

Eisenhower cared little f stance again had placed h bodied. As both a tool a and the opportunity of component of American culture after World War II. America's high-energy existence contained many other significant planks, including new patterns of living, shopping, and eating; however, the automobile is the device most responsible for the mindless spread of expansive energy use in the lives of grassroots consumers after 1945. Even without the policies crafted by Eisenhower, the automobile as a consumer item emerged before World War II and then synched with the critical emergence of the middle class and conspicuous consumption to solidify a clear postwar transition to an era reliant on petroleum consumption for everyday living.

During the years that surrounded and included the convoy, e energy transition can be seen as a grand, top-down orchestra- corporate interests and government. After 1945, though, utomobile a consumer device was accomplished not cturers, advertisers, and government cooperation. were also a big part of the equation. Americans fact, we wanted more than one. A passion for hicle performance and appearance to emerge rest for many Americans, particularly men. and cost soon also became a demarcation g within society. Indeed, by the 1950s, mo- ke an entirely different device from Henry able only in black. Clearly, this was part of hat Eisenhower observed as well, particu- awareness of how transformative effective an entire culture.[197] larly of course, conspicuous consumers of the les and their expectations grew rapidly. Big Three auto manufacturers (Chrysler, ted against one another through the styl- By making available new accessories and to manufacturers created a hierarchy of lines. In trying to outdo each other, the

tion by making the a only by manufa Now consumer wanted a car—in driving allowed ve as a new leisure fe And vehicle for economi tor vehicle Ford's M the en

Big Three also began cycles of review and improvement that led each one to establish annual schedules for modifying their lines. This shift in the early 1950s had two tremendously important outcomes: first, it put out of business the smaller manufacturers who could not afford to keep up with the Big Three; and second, it introduced the concept of "planned obsolescence" to the auto industry and to American consumerism in general (inspiring consumers to purchase new models as opposed to last year's). The simplicity of the convoy's emphasis on Packards in 1919, for instance, had now swelled into a dynamic, competitive market-place. From the "seedling" era of the early 1900s, the American automobile had now reached full germination to become much more than a transportation device.

Between 1945 and 1955, passenger car registrations in the United States doubled from 25.8 million to 52.1 million. With these vehicles came a whole host of sociological connections and amplifications that formed an entire era of intense consumption. Auto manufacturers were among the first to master this new universe of consumption. Design historian David Gartman calls the automobile "the central icon in this spectacle of consumption." In 1955, Americans bought a record 7.9 million cars, which marked 19 percent more than the previous best sales year in 1950. Ninety-five percent were made by the Big Three.[198]

These postwar trends were operated with mastery by business titans, much as Ford had performed decades before. In particular, the corporate giant Alfred P. Sloan merits credit for making Americans relentlessly desire new autos. By institutionalizing auto shows and the release of new models annually, Sloan created an artificial cycle—similar to a season within nature—that influenced the lives of many Americans, whether they manufactured, purchased, or merely lusted after new automobiles. Building off the World's Fair spectacle, Sloan created during the 1950s an annual extravaganza known as the

General Motors Motorama. Demonstrating his understanding of how the emerging cultural systems of the United States functioned, Sloan made these shows into media events and invited the financial elite. Auto designers picked up cues of modernism seen in airplanes, railroads, and buildings at the time. They also found inspiration in the culture-wide obsession with space travel.[199]

GM led the way with innovative designs, and consumers flocked to buy their cars. Ford and Chrysler struggled to invigorate their design departments to keep up with GM. Thanks to this competitive marketplace of the 1950s, engineering took a backseat to styling at each of the Big Three. Gartman writes that by the end of the 1950s, GM had institutionalized styling, with internal policies that prioritized "new model development, car line-ups, and planned obsolescence." Leading the way was Cadillac, the first brand to achieve clear symbolism as an elite product that embodied success—as Packard had decades prior. GM actively cultivated the brand's unique mystique.[200]

Even the designers admitted that they overdid it in the 1950s. The decadence was symbolized by Ford's Edsel. With its concave scalloped sides and gull-like horizontal taillights that countered the vertical fins on most designs, teeming with automatic gadgets for all passengers, and powered by one of the most powerful engines on the road, the Edsel represents for Gartman "the epitome of the automotive excesses of the decade." Importantly, Ford also made sure that the Edsel's 1957 release set a new standard in terms of advertising and hype. Regardless, the Edsel was a commercial flop. Critics called it unnecessarily overdone and focused on its vertically pronounced grill, which distinguished it from any other vehicle. Designers had assumed that consumers wanted these features and wished to be different from others, but they overshot in this case. By the end of the decade, auto manufacturers had learned that the consumer mind was not quite so simple, even if the automobile was wound into living patterns more than at any previous time.[201]

As Eisenhower emerged as a leader of 1950s America, there was no apologizing for such excesses. Petroleum, in particular, was the hidden hand, entirely overlooked in the growing obsession over vehicles and roads. Its sufficient supply may have been the single greatest assumption behind American exceptionalism of the era. Americans, in particular, invited petroleum into nearly every aspect of our lives during the late twentieth century, from the flexible form of disposable plastics to the chemical basis for pharmaceuticals, and much more. Its affordability during most of this era made a reliance on petroleum seem sensible.

Eisenhower was masterfully aware of the importance of consumers in the public sphere of 1950s America. As we note the excesses of consumption during this energy-intensive era today, though, it is crucial to recall that the status of petroleum never changed: even when its supply made it cheap enough to use in mundane activities such as manufacturing toothpaste tubes, oil was a finite resource—it would run out. By ignoring this reality, though, American society significantly altered human existence, and our standard of living became the envy of the world. Although the results of this new ecology of human life would prove terribly damaging, the post–World War II era of conspicuous consumption marked the blissful expansionist time of no questions. If, instead of the results that would become evident by the end of the century, we focus our view on how our living environment was altered, we see the fundamental fashion in which inexpensive oil changed the way humans live in the world. Food, shelter, mobility—each one basic planks of our lives—became more widely available and accessible in the late twentieth century. For momentary convenience, comfort, and ease, we chose a standard of living that could not persist indefinitely.[202]

Similar to previous eras when Thomas Jefferson identified American character with the sturdy virtue of those who worked the land, Eisenhower located the basic traits of the American era of conspicuous consumption within our automobility. Instead of emphasizing new features of auto design, though, his insight was on the systems required to apply the new technology.

Gartman writes, "The explosion of suburbanized consumption that contained the larger ambitions of working Americans depended on expanded automobility, not only for physical transportation to suburbia but also for cultural insulation and gratification within it." When manufacturers sought to exploit the public's interest in the 1950s, the auto industry "blazed a path of fantastic, escapist consumption."[203]

These impulses were at least part of the explanation for Americans' choice to expand its roads after World War II; however, railroads also expanded significantly. Historian Jacques Barzun described railroads as an institution and "a source of poetry," but they did not inspire Americans in the 1950s. Driving restrictions during the war had increased ridership on rails, which would after the war give way to degraded rail lines and companies heading toward bankruptcy. Even if Americans had approached railroad development with more committed resources, it is doubtful that they could have competed with the rising consumer passion for the open road: rail's rigid structure seemed a stark, immobile—old-fashioned—contrast to the flexible modernity that the automobile offered.[204]

THE AMERICAN model of expansion in this area grew out of the modes of transport that had been introduced in the convoy of 1919, foreshadowing how vehicles might be applied to an ideal lifestyle. Factored in with the post-war need for housing, the new American living environment assumed auto travel. Historian

Lizabeth Cohen describes this trend as the construction of a "Consumers' Republic" in which consumption was construed as an expression of national identity and even patriotism. She writes, "This spreading affluence in spite of extensive market controls fed business leaders' and government officials' optimism that 'a peacetime economy of abundance' would follow the war, as the president of Studebaker Automobile Company assured a National Association of Manufacturers meeting in 1944." After World War II, Cohen traces the emergence of "a new postwar ideal of the purchaser as citizen who simultaneously fulfilled personal desire and civic obligation by consuming."[205]

Cheap energy, though, was almost entirely unseen in the background. By 1920, most American planners and developers created a transportation infrastructure upon the assumption that consumers would be using cars to move about. Even home designs came to be based entirely on this assumption. Outside of major cities, the result became the American suburb. While the first suburbs followed railroad and streetcar corridors, more affordable automobiles paved the way to a more singular vision of the American Dream: owning one's own, single-family house on a large lot in the suburbs. This trend fell in sync with many consumers' desire to leave the city in order to raise children in a safe, clean environment.

Post–World War II America consumer tastes drove planners to design homes and entire developments linked to other services by only the automobile. The planning system that supported this residential world depended on the existence of reliable roads as well as an intricate mechanical and cultural infrastructure. New outlying suburbs each needed to also be integrated into larger designs by planners, until most of the U.S. seemed linked through the use of automobiles. The demographic shift is clear: 33 percent of the nation lived in urban areas in 1950, 23 percent in suburban, and 44 percent in rural; by 2000 over 50 percent of

Americans lived in suburbs. This trend is described by historian Clay McShane in this fashion: "In their headlong search for modernity through mobility, American urbanites made a decision to destroy the living environments of nineteenth-century neighborhoods by converting their gathering places into traffic jams, their playgrounds into motorways, and their shopping places into elongated parking lots."[206]

After World War II, most planners simply assumed that residents of such an environment sat behind the wheel of a car. Instead of the Main Street prototype of small-town America, a new model emerged that brought shopping and other services out to the suburbs and to consumers. American planners such as Jesse Clyde Nichols initially devised these type of shopping areas as a hybrid of previous forms and a site such as Kansas City's Country Club District offered services to consumers who drove there. These were just a step, though, in the evolution of the "strip" that made services available in almost every community with just a bit of driving. Clearly, though, these areas integrated the car into the design program.

This transformation of the American landscape did not occur by chance or by any great action on the part of Eisenhower. Through the work of trade organizations (such as the Outdoor Advertising Association of America and the Automobile Association of America after 1940) and other interests, zoning laws and regulations became tools for the advancement of this architecture of mobility from the 1930s forward. Few questioned this construction, because the dramatic shift in convenience appeared to be indisputably progressive. As a common good, this convenience, as with the overall use of the automobile, became a primary tool for raising the American standard of living and defining and expanding the middle class during the twentieth century. The corridors of consumption were paved with asphalt, called roads, and generally funded by taxpayers. Once organized, established passages allowed gas-powered vehicles to become conceivable; as these roadways became more practical, entrepreneurs and government officials helped to make them essential. And Ike's experiences earlier in life made him the perfect leader for this moment.[207]

IN THE wake of the 1919 convoy, America began to set the model for a society that confidently funded infrastructure. Although some communities installed toll roads in the early decades of the 1900s, most others relied on general taxation to support transportation. The convoy had been a dramatic illustration of this need for public attention and funding. This emphasis, writes historian Kenneth Jackson, "is testimony both to the public perception of the benefits of automobility and to the intervention of special interest groups."[208]

During the next few decades, progressive communities used tax dollars to purchase the means with which to create their own asphalt roadways. Telford machines and other variations used cheap petroleum to create a longer-lasting tar surface that would take existing paths and roads made of packed dirt and cordwood planks (often referred to as corduroy roads) into permanent passages for independent travelers. Jackson emphasizes the connection between the earliest suburbs and the first expressways (or parkways) that helped to stabilize and systematize the consistent reliance on auto travel to work. This tradition dates to the New York area and particularly the Bronx River Parkway, on which work began in 1906. By 1930, the northern suburbs of New York City, primarily in Westchester County, were connected to the city by a system of expressways. Commuters could continue to travel by rail, but now they could also opt for the more flexible independence of auto commuting.[209]

Nationally, urban spokespeople joined forces with farm groups to demand a national highway program, and the early efforts on behalf of the Lincoln Highway marked just such an effort. Prior to the convoy, the Federal Aid Road Act of 1916 offered funds to states that established highway departments. With the wild success of the convoy, the act was updated in 1921 to list two hundred thousand miles of existing road as "primary," making them eligible for matching federal funds for development and expansion. In addition, the updated act created a federal Bureau of Public Roads that was empowered to create a highway network that would connect cities with populations over fifty thousand. Throughout the pre–World War II years, transportation development was organized by a confounding logic: consistently, the road was defined as a public good that would spur additional ancillary development and, therefore, deserving of public funds; by contrast, mass transit was most often viewed as a private business that was unworthy of aid.[210]

For the nation, the sway of transportation had certainly swung in the direction of the automobile through stunts such as the convoy of 1919; for Eisenhower, personally, his commitment to roads still remained dormant as he focused on the needs of the World War II effort. As industries tied to automobility grew in prominence over these decades, they often used their power to squelch out potential competitors. For instance, from 1926 and deep into the 1950s, GM operated a subsidiary corporation tasked to purchase streetcar systems that were in financial difficulty and to retrofit the systems to run on rubber tires. This retrofitting required the removal of track systems, an expensive process in its own right. In cities such as Los Angeles, the tracks were removed and the routes run by GM-manufactured buses. Although GM was found guilty of criminal conspiracy in such cases, it was fined only $5,000. Jackson writes, "The misguided and unfortunate result of such thinking was that Americans would no longer have transit options and that the car would become a prerequisite to survival, with disastrous consequences for energy consumption and traffic deaths." Eisenhower never expressed such a critical view, and he never perceived the need for moderating our growing dependence on petroleum. For the good of our society, in fact, he sought expansion so that as many Americans as possible could have access to a higher standard of living.[211]

During the post–World War II era of massive home construction and suburbanization, federal and local subsidies also spurred the scattered construction of roads. Between 1945 and 1954, nine million Americans became entirely reliant on roadways for their everyday life and moved to suburbs. Central city populations grew by just 10 million between 1950 and 1976 while suburban growth was by 85 million. The decentralization of the population marked a remarkable opportunity for building and expansion as communities seized the opportunity to develop the arteries connecting suburbs to cities.[212]

In a world designed around consumption, shopping malls became the new marketplace. The smaller variety, strip malls, connected roadways and parking lots, in a form that could be placed nearer the suburb. Together, the composite architecture of mobility is now described with a disdainful term: "sprawl." Incorporating suburbs into such development plans, designs for these pseudo-communities were held together by the automobile. Sprawl oozed out from these starting points. Defined by low population density, this pattern of design had unique appeal for Americans for two reasons: it could be organized not by demographics, ethnicity, or economic class but instead by consumptive patterns—and it required large amounts of space. As the United States became a nation of consumers after World War II, sprawl became home to most Americans as we worked to fill our vast amounts of open land for development.[213]

Roads appeared to be a clear strategic opportunity for development, but primarily at the municipal and state level. The process for road-building typically proceeded as follows: states planned road networks, raised the necessary money through bonds, taxes, and license fees, and then built roads with the purpose of serving local communities. The postwar America that Eisenhower viewed was one defined by a lack of logic and coherence where these locally developed systems came together. Efforts to bypass congested small-town roads led to plans such as the Federal-Aid Highway Act of 1944; however, with questions of funding, Congress often still balked and did not approve such daunting projects. The stage was clearly set for a leader who would make federalizing roads a national issue.

EISENHOWER'S WILLINGNESS to think about roads on the national level began with the convoy, but expanded with his other experiences during the World War II era. Like the United States, other developed nations had impulses to modernize, most famously in Germany under Hitler's ruthless rule. The centralized control of a more authoritarian government allowed construction of the Autobahn to begin in the late 1920s, and when Hitler assumed power as chancellor of the Third Reich in 1933, he made its expansion a symbol of his effectiveness and control—well in advance of the widespread ownership of automobiles that might justify such an expense. When wartime needs halted its construction at the end of 1941, 2,400 miles had been completed and 1,500 more were under construction. Earlier, in a national effort to fill the miles of road with a small, affordable "people's car," Ferdinand Porsche designed the Volkswagen in 1938. Over 360,000 Germans paid in full or in installments for the vehicle in advance of its production. However, in August 1939, Hitler ordered Porsche to switch the Wolfsburg plant to production of military vehicles based on the Volkswagen.[214]

Under the leadership of Eisenhower, by the time the Allied forces reached Germany, they could, ironically, also take full advantage of the autobahn. "After crossing the Rhine and getting into the areas of Germany served by the Autobahn," recalled E. F. Koch, a US Public Roads Administration employee, of Allied movement in 1945, "our maintenance difficulties were over. Nearly all through traffic used the Autobahn and no maintenance on that system was required." Throughout the Allied pursuit of German forces across Germany, the Autobahn proved invaluable in moving supplies into place behind troops. As an integral part of the planning for these Allied maneuvers in Germany, Eisenhower carefully studied the entire German system of roads. Different from the US system, this was a rural network that largely left German's major cities alone. The clearest distinction, though, was that Germany's model was to first build roads and then cars would follow; the United States had proceeded opposite. In his wartime experience, though, Eisenhower had acquired a uniquely international insight, informed by the strategic foresight implemented in nations with centralized government and authority over expansion.[215]

With cars already clogging cities, American road planners Thomas H. MacDonald and Herbert Fairbank of the US Public Roads Administration (the name of the Federal Highway Administration's predecessor during the 1940s) argued for a different approach. Unlike in Germany, traffic volumes were high in America, where car ownership was widespread. Congestion in America's cities had long been a serious complaint that MacDonald and Fairbank would address in their vision of the Interstate System.

To Eisenhower, the basic commitment to a system of organized roads was a lasting model of modern advancement. "During World War II," he recalled, "I had seen the superlative system of German autobahn—[the] national highways crossing that country." "After seeing the autobahns of modern Germany and knowing the asset those highways were to the Germans, I decided, as President, to put an emphasis on this kind of road building. . . . The old [1919] convoy had started me thinking about good, two-lane highways, but Germany had made me see the wisdom of broader ribbons across the land." Although his commitment to the vision of systematic roads was clear, the reality of achieving this goal would be one of the heaviest domestic political lifts of his presidency.[216]

WHEN EISENHOWER stepped onto the political scene of this modernizing nation, his presence was significantly larger than that of most candidates for office. After retiring from the military, he spent two years as president of Columbia University and had constant flirtations with representatives of the Republican Party about running for president. Initially, Ike's popularity drew from his standing as a military hero; however, by the time he ran for president in 1952, he had designed an ideological approach with wide appeal. These factors combined with growing anxiety about the Cold War to transform Ike into a wildly popular, larger-than-life figure during the 1950s. Indeed, he remained middle-of-the-road on nearly all political issues and polled as the most popular human on earth throughout the 1950s—even while he served as US president until 1961. Historian and biographer William I. Hitchcock writes, "The last president born in the nineteenth century welcomed innovation, modernity, new technologies, the space age, and global travel and communication."[217]

A spirit of thoughtful compromise predominated Eisenhower's ideological approach to leadership. It was ultimately termed the "middle way," a moderate, almost apolitical approach. According to Hitchcock,

> The key to Eisenhower's success lay in his ability to balance, in his own person and in his policies, the contradictions in American society. . . . [Journalist William Lee Miller in

1958 observed that Eisenhower] found a way to reconcile the cross-cutting tendencies of the American character: the "practical, competitive, individualistic, externally minded, environment-mastering and success-seeking on the one side, and the spiritual, idealistic, friendly, team-working, moralizing, and reform-seeking on the other. Mr. Eisenhower exactly summarized both."[218]

In hindsight, it is clear that this ideology provided Americans with significant comfort and security in a troubling time. It was a vision of America in which citizens of the middle class felt that much was accessible and that with a bit of care, they could have it all—as individuals and collectively as a society. By 1950, the average American family income was $4,237, far exceeding that of the Depression era. By 1960 it had risen to $6,691. Home ownership rose from 48 percent to 53 percent. Car ownership from 59 percent to 73 percent. Americans in the 1950s enjoyed a standard of living higher than that of any previous generation.[219]

While Eisenhower did not cause this prosperity, his policies did make a significant difference. Uniquely above the fray of everyday debate and politics, he made choices that reflected his philosophy of government. He described his politics as "a liberal attitude toward the welfare of people and conservative approach to the use of their money." With this luxury during the 1950s, Ike could pursue large-scale initiatives that might intimidate or put off other politicians. To his mind, Republicans of the past had failed to use the government to accomplish public good, to be active and creative. He felt that President Herbert Hoover, in particular, had failed to act vigorously to address the desperate economic conditions leading up to the Great Depression. Eisenhower wanted to find policies that allowed for a proper role for government intervention in the economy while encouraging individual initiative. As Hitchcock puts it, "Eisenhower was not a small-government conservative, although he successfully sold himself as one to the public. He believed government should create the conditions in which Americans could pursue their own ambitions." As he considered the choices in front of him, Ike's ideology and personal experiences, particularly the 1919 convoy and his use of the Autobahn in 1945 Germany, pulled a leading cause into focus.[220]

ROADS BECAME a top priority for Ike as he shifted into politics; however, his unwavering commitment did not translate into a simple political path forward—making the sausage, as always, proved ugly.

After his landslide victory over Democrat Adlai Stevenson for US President in 1952, it remained unclear exactly what type of transportation initiative he would propose. To negotiate the complexities of planning and implementing a national project of such scope, Eisenhower tapped one of his most trusted advisors: Lieutenant General Lucius Clay. A West Point–trained engineer who had worked closely with Ike throughout World War II (and particularly on the D-Day logistical planning), Clay was appointed to head the President's Advisory Committee on a National Highway System, which would eventually be known as the "Clay Committee."

While the early planning and work went on behind the scenes, the situation was complicated by Eisenhower's health. In fall of 1955, he had a heart attack that weakened him for six weeks, and in June 1956, he underwent an operation for a bowel obstruction for which he was in recovery for four weeks more. Although the Republicans controlled both houses of Congress during this time, few observers expected that the president would want to prioritize a huge highway bill while his own situation appeared quite tenuous.[221]

Behind the scenes, Clay proceeded with the same sort of team-work that he had employed in the planning and execution of the D-Day invasion. Unlike F. D. R., who had constructed maps to show MacDonald where to place roadways, Eisenhower left the details to others. He simply made the orders, the first of which arrived to his assistants on April 12, 1954, when he told them that he wanted a "dramatic plan to get $50 billion worth of self-liquidating highways under construction." The committee began meeting in October 1954 and submitted its report in January 1955.[222]

Although Clay had aided Eisenhower's stoic chief of staff, Sherman Adams, the former governor of New Hampshire, in making most of the president's cabinet appointments, he had since returned to the private sector as president of the Continental Can Company, and he held little tolerance for Washington politics. To join him, Clay chose four well-positioned leaders of business and engineering firms: Stephen D. Bechtel, chairman of the board of the Bechtel Corporation; S. Sloan Colt, head of Bankers Trust; William A. Roberts, head of Allis-Chalmers, which manufactured massive construction machinery; and David Beck, head of the International Brotherhood of Teamsters.

It is this group and this moment, which many conspiracy theorists point to as the "General Motors Cabinet," that made sure that America turned away from mass transit and toward gasoline-powered, individualized transportation. Very likely, the facts are true; however, this strategic commitment is a vital portion of what propelled the United States to the head of the era of high-energy consumption. Trolley companies in many American cities were already experiencing financial hardship, and the accomplishment of Eisenhower's assignment proved the death knell for many of them. It is just such a moment of choice by leadership that reflects a clear push to extend and to institutionalize the transition. Ike's thinking was unequivocal.[223]

The reality was that the resulting report to Congress on the National Highway Program was organized around a "Grand Plan" that called for $50 billion of federal money over a span of ten years to construct a "vast system of interconnected highways." The committee's proposal was organized around four points, starting with safety. In its findings, it specified that accidents resulted in thirty-six thousand traffic fatalities each year and the accompanying multibillion-dollar effect of such deaths on the economy. In terms of economic impacts of bad roads, the report also cited that the poor physical conditions of existing roads cost owners thousands of dollars in repair. While these points might be true at any point in history, the report's third point emphasized a unique concern of the moment: national security. Historian Lee Lacy writes, "The pervasive threat of nuclear attack in the United States called for the ability to execute the emergency evacuation of large cities and the quick movement of troops essential to national defense." Finally, the plan appealed to the idea of economic growth for the nation and argued, according to Lacy, that "improvements in transportation must keep up with the expected increase in US population. Moreover, road improvement was essential to the economy and an efficient use of taxpayer money." In totality, according to Lacy, the Clay Committee concluded that "the positive economic attributes of the highway system were the potential for economic growth and the well-being of the economy through 'speedy, safe, transcontinental travel' that could improve 'farm-to-market movement.'"[224]

Thus, Clay and his committee clearly saw their work as a project of engineering and planning—not politics. But Clay despised the work of the political world. In fact, it is reported that in the month that the committee completed its work, Clay's tab at the White House Mess was only 65 cents.[225]

The implementation of Clay's plan was entrusted to a team of political operatives. Heading the Bureau of Public Roads,

Francis du Pont, son of the famous chemical-fortune family, had built a reputation in developing roads in his home state of Delaware. Sherman Adams proposed a "Continental Highway Finance Corporation" under the departments of commerce and defense that would see to the financing of the highway system. And Eisenhower appointed Major General John Stewart Bragdon (who had overseen massive construction projects for the military in World War II) as head of the Public Works Planning Unit of his Council of Economic Advisers. From that post, Bragdon proposed creating a "National Highway Authority" that would abolish state highway agencies. Ike was not afraid to create political waves if an idea made sense to him!

Unintentionally, through this issue, Eisenhower had reopened a historic and ongoing debate in the United States over states' rights and the amount of federal control that should be accepted. Even Thomas Jefferson, who presided over significant accomplishments of federal authority, such as canals and rivers, allowed the National Road to be built only because it was paid for with revenue from western land sales. He believed the construction of roads that crossed state boundaries would require a constitutional amendment. When Vice President Nixon presented Eisenhower's plan to the Governors' Conference in 1954, the idea created a great buzz, particularly if it drew funding from gasoline taxes. But critics still feared the overreach of the federal government into local affairs as essential as road placement and design. The trump card to assuage these concerns, though, grew out of the unique spirit of panic that undergirded all of American life at the dawn of the Cold War.[226]

Indeed, the single most likely reason for the success of the highway project was its connection to national defense at the exact moment of the emergence of the Cold War, which would define so much of life, diplomacy, and politics in coming decades. In the mid-1950s, the public was deeply worried about the nation's emerging nemesis, the USSR. In most cities, for instance, civil defense coordinators found spaces that could serve as shelters in case of nuclear attack. Many communities tested their air-raid sirens at noon each day. And, of course, in many public schools, children performed "duck and cover drills" similar to fire drills.

In such an anxious environment, the best form of preparedness was to enable citizens to flee the population centers that were among the nation's 185 most likely targets of attack. For instance, in his 1954 speech to the governors, Nixon used the words "atomic" or "atomic war" at least ten times. Were nuclear war to break out, 70 million urban residents would have to evacuate by road. The threat of nuclear war also provided logic to the Clay Committee's call for roads that would enable large-scale evacuation from most major cities. The Committee, writes Lacy, soberly stated in its report, "The rapid improvement of the complete 40,000-mile interstate system, including the necessary urban connections thereto, is therefore vital as a civil-defense measure."[227]

Along these lines, in June 1955 a large-scale urban evacuation drill highlighted the urgent need for a national evacuation plan. Similar to the 1919 convoy, this drill reinforced the need for change when existing roads and systems proved inadequate. The overcrowded evacuation roots and overall confusion only added fuel to the call for highways as they related to national security. And, indeed, early on the administration underscored the role of a uniform system of roads for national defense and ordered the Department of Defense (DOD) to get involved. Lacy reports that this coordinated effort was reinforced by the construction of a central Illinois testing facility that used DOD-contributed equipment and personnel to evaluate pavement, road standards, and construction techniques. The construction of roads had clearly gained even more importance from the intensifying Cold War.[228]

The two previous world wars had convinced military leaders that roads were vital to national defense. As part of Cold War preparedness, over two year of testing, Lacy reports, Army trucks logged as many as 17 million miles on the test roads. As a result of the tests, standards for highway construction and maintenance were developed that would withstand military and civilian needs and provide the necessary passages throughout the nation. Clearly, such demonstrations assuaged anxiety; however, as with many details of the Cold War effort, they also reinforced the deep and constant concern for public safety and preparedness.[229]

HOW DID the United States get the most advanced highway system on earth? Certainly, it is complicated to answer this question. The simplest response, for that reason, may be Ike. However, even though the convergence of this moment was clearly initiated by Eisenhower, his medical travails left him even more fervently committed to finding an accommodation with all the disparate political perspectives well into 1956. In this spirit, he signed a bill that even he—with his passion for road-building—had initially resisted.

Paying the $50 billion price tag was the real source of consternation, of course. From where would the funds be drawn? Already by 1954, Washington had become a web of complicated interests. On the topic of highways, these included the Association of General Contractors, the National Asphalt Pavement Association, the American Concrete Paving Association, rubber manufacturers, automakers, truckers' associations, and on and on. As Clay weighed proposals for federal toll roads and new taxes, the plan instead argued for financing the project through bonds. A new national corporation created by Congress would sell thirty-year bonds to finance road construction, of which the federal government would repay ninety percent from the federal tax on gasoline and diesel fuel. States would share the remaining ten percent of the debt. Selling bonds would make the money available immediately, they wagered. The congressional proposal would cost $101 billion over ten years for primary, secondary, urban, and rural highways and an Interstate Highway System (IHS). Congressional representatives believed that Eisenhower would approve because it would raise neither taxes nor the debt.[230]

When Congress convened in January 1955, the White House watched helplessly as the Senate Finance Committee refused the idea of bonds, which Virginia Senator Henry Byrd argued "would be the end of honest bookkeeping." Tennessee Senator Albert Gore Sr. drafted an alternative bill that called for the Bureau of Public Roads to spend $10 billion on an interstate system through 1961. Other senators' proposals followed, and soon the IHS was dead for 1955. In fact, extended debate over funding brought the Clay Plan into deep existential peril.[231]

In the interim, Eisenhower suffered his serious heart attack while visiting friends in Denver, Colorado. The experience and his long recovery period left Ike even more convinced that at all costs he had to make highways happen for the future development of the nation.

When the identical bill was submitted in 1956, the method of financing was left to the House Ways and Means Committee and Maryland Representative George Hyde Fallon. To pay for the system, he introduced the Highway Revenue Act, which would pay into a separate Highway Trust Fund and thereby keep the expenditure out of the federal budget. Referred to as a "pay-as-you-go basis," the Trust Fund appeased most of the critics, and the administration abandoned its plan to use bonds. On June 25, both houses of Congress approved the conference bill. In addition, the bureau had shrewdly designated many additional miles of interstate for cities and thereby broadened the range of interested parties. A swell of support brought the Federal-Aid Highway Act of 1956 before Eisenhower for his signature on June 29; he signed the bill at Walter Reed Army Medical Center, where he was still recovering from bowel surgery. Even though its planning came so intimately from his personal experiences and choices, and its impact would be felt so broadly, the highway bill was only one of twenty pieces of legislation Ike signed that day.[232]

"YOUR HIGHWAY Taxes at Work" was the sign that met frustrated drivers in Virginia in spring 1959. Among the vehicles slowed by the road construction, ironically, was the presidential motorcade that carried Eisenhower to Camp David, the presidential retreat named for Ike's grandson.[233]

The system of roads that would define the nation took decades to construct and required a great deal of additional debate and discussion, but Eisenhower had started the massive ball rolling. While other societies in the world invested in transportation networks that diverged from this model, the United States built roads—and lots of them. Roads and highways, of course, were only useful if they could be populated with drivers and their vehicles. In the process of overseeing the national construction of the world's leading system of roads, Eisenhower had unquestioningly wedded the country to the high-energy era. His choice, for the American people, had been a most simple, functional one.

EPILOGUE

Choosing Change

Critical lessons need to be drawn from Ike's 1919 road trip. However, the most critical of all might be the simplest historical lesson contained in these pages: *this energy transition happened.*

Energy transitions do occur, and in 1919 Eisenhower was living through one of the most significant in human history. The study of history allows us to start from this fact and then to piece together the components that emerge as important to the occurrence of the 1919 convoy. At the center of this drama is the young man from Kansas. Behind him, Americans chose a path of energy consumption that was expansive, multiplying, and ultimately finite. Our legacy of the high-energy period—and of the accomplishments of Joy, Ford, Clay, and Eisenhower—can be a proud one. But it can't overlook that last reality while remaining honest. The transition to fossil fuels was and remains a model of success destined to end. And, if we organize our understanding of the 1919 convoy in this fashion, our energy transition in 2024 becomes smarter and smoother.

In 1919, the converging factors around the convoy were truly unique. Business and military interests combined to stimulate national growth. Publicity for the convoy did not emphasize energy, but a vision of a successful American society had been pitched by many leaders and politicians in the early twentieth century—and that vision was built on a massive supply of petroleum. This required a complex transition on every level of human life. Anxiety and fear—indeed, even practical, systematic planning— opposed this massive shift in society. What would we reliably drive? Where would we park such a vehicle? How would we refuel when needed? Where would we live? How would we shop for our daily needs? To every one of these questions and a host of others, Americans devised thoughtful, ingenious responses. Often, those enacting change profited mightily, constructing an entire economy that, like falling dominoes, had a cascading effect.

For some, great changes began with seemingly insignificant choices. Ike's personal choices began with a willingness to venture from the Kansas prairie and into the military. Each decision that followed, including joining the convoy, was informed by his experience and his view of the changing, modern world that surrounded him. Behind each decision, too, was a confidence that Americans would step into the breach. Eventually, Ike would become a key player in America's massive energy transition.

During the energy transition through which he lived, Ike led soldiers to discover ways to alter battlefield tactics and strategy with the use of petroleum. The tank was an emerging wartime technology in the twentieth century and Eisenhower applied it immediately in 1918. Decades later, he was a military leader at a time when success on the battlefield could be directly tied to the planning and infrastructure necessary to maintain a steady supply of crude. And, finally, as the Cold War emerged as a conflict over technology and lifestyle, Ike guided the nation to unabashedly leap into the high-energy era and to tie its economic development to a federally subsidized system of roads that would host gasoline-powered autos. When we look at these efforts throughout Ike's lifetime, in addition to his participation in the convoy, we see the type of decision-making that spurs, extends, and defines an energy transition.

In the half century prior to 1950, humans doubled their global energy use. In the next half century, humans multiplied their global energy use *by five times*. J. R. McNeill and Peter Engelke explain the fundamental shift in their book *The Great Acceleration*: "The creation and spread of fossil fuel society was the most environmentally consequential development of modern times. . . . It enabled many activities that otherwise would have been uneconomic and would not have happened, or perhaps would have happened but only much more slowly."[234]

Powered by the great acceleration of energy use post-1950, living patterns became so impactful that scholars refer to the current period of human life as the Anthropocene, the geological epoch when human activity became the determining factor in earth's future. Of this era, McNeill and Engelke write, "Cheap energy gave people new leverage with which to accomplish things, move fast and far, make money, and, if inadvertently and often unknowingly, alter the environment. Almost everyone who could take advantage of cheap energy did so." Just as Eisenhower and others observed the potential of expansive energy supplies prior to 1950, today's energy transition offers its own unique promise. In a fascinating twist, an important part of this opportunity is the basic realization that the high-energy era's assumption of limitless energy supplies was faulty and endangered our living environment at its foundational level. Our new understanding promises a future of sustainability that was never considered in 1919 as the convoy blew through the United States to demonstrate a new way that Americans might define progress. The common theme between these transitions, though, is a willingness to embrace radical societal change.[235]

Emphasizing the radical, some critics of the high-energy era today propose actions to sabotage our energy infrastructure—a way of jarring the system out of its comfortable state of efficiently supporting our everyday way of life. Certainly at odds with the choices made by Eisenhower and others (that is the point), radical action misses the most profound opportunity of our energy transition.

In 1919, even the basic role of every American as an energy consumer was new. Options were few, and most of the consequences of the high-energy era emanated from each of us simply living life as the energy culture took shape around us, guided by infrastructural choices made by leaders such as Eisenhower. Cheap energy made America go. We built our homes and heated and electrified them while also faithfully powering our vehicles just as society guided us. Eventually, we complained when the cost of performing our essential energy uses rose too high. Otherwise, part of being a good citizen was simply to participate in this energy economy that undergirded the nation.

Today's energy marketplace is different. In a fracturing and diversifying energy market, consumers participate in a great transition just by making our daily choices. Armed with a century's worth of scientific knowledge, options bring us a degree of control over our consumptive patterns never imagined in 1919. Unlike the great wave of change that put a century of Americans behind the wheels of ICE-powered autos, consumers today enjoy a marketplace of various viable options. As consumers, we each can do our research and be part of a sea change of electric vehicles, solar panels, and home energy storage devices. Our choices fuel today's energy shift and force the industry to keep up.

In 2024 it is not difficult for us to imagine the anxieties spurred by the massive disruptions in the 1910s. As Ike knew so well, government's role was to observe these trends and to foster methods for expansion and opportunity that grew from them. Leaders needed to deftly observe their times and not to ignore the undeniable trends. Instead, their task was to help to create a society that could benefit and further the trends that were playing out. Ike was cognizant of massive supplies of raw resources that would generate power for human living; however, instead of purchasing stock in energy companies that would derive personal profit for himself and his family for years to come, he led the country in ways that would prove advantageous for the greater good. Indeed, his choices fostered an energy transition that he believed would benefit the nation.

Finite energy supplies were the fuel for American success. Almost organically, Ike merely left himself open to his moment, joining the 1919 convoy, and to moments that followed. His choices helped to guide Americans toward a lifestyle that was more energy intensive than any in human history. A nation ascendant, we led our species to longer, more productive models of living while soiling earth's future in potentially catastrophic ways.

Our new transition grows from a nuanced awareness of energy supplies and the impact that each reaps on the earth. But our modes of living still undeniably revolve around energy. That part of the 1919 convoy has continued to move forward. While we continue to be an energy intensive society, the opportunities for growth in our transition clearly lie in the nuance: deriving energy more cleanly and more sustainably; understanding better the processes of how we get it and use it; and managing and conserving it with more care. In the background of the convoy of 1919 was the basic awareness—though normally unstated—that energy supplies were the twentieth century's gold; in 2024, we maintain this realization but accent it with a significant awareness that supplies are just the start. In today's transition, many actors are at play, making choices that might define a more promising future. But our moment's Ike likely has not yet emerged.

Bibliography

Allitt, Patrick. *A Climate of Crisis: America in the Age of Environmentalism.* New York: Penguin, 2014.

Ambrose, Stephen E. *Eisenhower: Soldier and President.* New York: Simon & Schuster, 1990.

Birkner, Michael J., and Carol A. Hegeman. *Eisenhower's Gettysburg Farm.* Charleston: Arcadia, 2017.

Black, Brian C. *Climate Change: An Encyclopedia of Science and History.* New York: ABC-Clio, 2013.

———. *Crude Reality: Petroleum in World History.* 2nd ed. New York: Rowman & Littlefield, 2020.

———. "Oil for Living: Petroleum and American Conspicuous Consumption." *Journal of American History* 99, no. 1 (2012): 40–50.

———. *Petrolia: The Landscape of America's First Oil Boom.* Baltimore: Johns Hopkins University Press, 2000.

———. *To Have and Have Not: Energy in World History.* New York: Rowman & Littlefield, 2022.

Black, Brian C., Karen R. Merrill, and Tyler Priest, eds. "Oil in American History: A Special Issue." *Journal of American History* 99, no. 1 (2012).

Black, Edwin. *Internal Combustion: How Corporations and Governments Addicted the World to Oil and Derailed the Alternatives.* New York: St. Martin's, 2006.

Blair, John M. *The Control of Oil.* New York: Pantheon Books, 1976.

Bliss, Carey S. *Auto Across America: A Bibliography of Transcontinental Automobile Travel, 1903–1940.* N.p.: Jenkins & Reese, 1982.

Bradsher, Keith. *High and Mighty: SUVs: The World's Most Dangerous Vehicles and How They Got That Way.* New York: PublicAffairs, 2002.

Brinkley, Douglas. *Wheels for the World: Henry Ford, His Company and a Century of Progress.* New York: Viking, 2003.

Bromley, Simon. *American Hegemony and World Oil: The Industry, the State System, and the World Economy.* University Park: Penn State University Press, 1991.

Burlingame, Roger. *Henry Ford.* New York: Random House, 1976.

al-Chalabi, Fadhil J. *OPEC and the International Oil Industry: A Changing Structure.* Oxford: Oxford University Press, 1980.

———. *OPEC at the Crossroads.* Oxford: Pergamon, 1989.

Chandler, Geoffrey. "The Innocence of Oil Companies." *Foreign Policy,* no. 27 (1977): 52–70.

Chastko, Paul. *Developing Alberta's Oil Sands: From Karl Clark to Kyoto.* Alberta: University of Calgary Press, 2007.

Chernow, Ron. *Titan: The Life of John D. Rockefeller Sr.* New York: Random House, 1998.

Chester, Edward W. *United States Oil Policy and Diplomacy: A Twentieth-Century Overview.* Westport, CT: Greenwood, 1983.

Churchill, Winston S. *The World Crisis.* 4 vols. New York: Charles Scribner's Sons, 1923–1929.

Clark, John G. *The Political Economy of World Energy: A Twentieth-Century Perspective.* Chapel Hill: University of North Carolina Press, 1990.

Cohen, Lizabeth. *A Consumers' Republic: The Politics of Mass Consumption in Postwar America.* New York: Vintage, 2003.

Coll, Steve. *Private Empire: ExxonMobil and American Power.* New York: Penguin, 2012.

Conard, Rebecca. "The Lincoln Highway in Greene County: Highway Politics, Local Initiative, and the Emerging Federal Highway System." *Annals of Iowa* 52, no. 4 (1993): 351–384.

Cook, Kevin L. "Ike's Road Trip." MHQ 13, no. 3 (2001): 68.

———. "Ike's Road Trip." HistoryNet.com. March 26, 2019. https://www.historynet.com/ikes-road-trip/.

Cordesman, Anthony H. *The Gulf and the West: Strategic Relations and Military Realities.* Boulder, CO: Westview, 1988.

Cote, Stephen C. *Oil and Nation: A History of Bolivia's Petroleum Sector.* Morgantown: West Virginia University Press, 2016.

Cowhey, Peter F. *The Problems of Plenty: Energy Policy and International Politics.* Berkeley: University of California Press, 1985.

Crosby, Alfred. *Children of the Sun*. New York: W. W. Norton, 2006.

Daly, Selena, Martina Salvante, and Vanda Wilcox, eds. *Landscapes of the First World War*. London: Palgrave, 2018.

Dann, Uriel, ed. *The Great Powers in the Middle East, 1919–1939*. New York and London: Holmes & Meier, 1988.

Davies, Pete. *American Road: The Story of an Epic Transcontinental Journey at the Dawn of the Motor Age*. New York: Henry Holt, 2002.

Denny, Ludwell. *We Fight for Oil*. 1928 Reprint. Westport, CT: Hyperion, 1976.

DeNovo, John. *American Interests and Policies in the Middle East, 1900–1939*. Minneapolis: University of Minnesota Press, 1963.

———. "The Movement for an Aggressive American Oil Policy Abroad, 1918–1920." *American Historical Review* 61, no. 4 (1956): 854–76.

———. "Petroleum and the United States Navy Before World War I." *Mississippi Valley Historical Review* 41, no. 4 (1955): 641–56.

Doyle, Jack. *Taken for a Ride: Detroit's Big Three and the Politics of Air Pollution*. New York: Four Walls Eight Windows, 2000.

Eckes, Alfred E., Jr. *The United States and the Global Struggle for Minerals*. Austin: University of Texas Press, 1979.

Eddy, William A. *FDR Meets Ibn Saud*. New York: American Friends of the Middle East, 1954.

Eisenhower, Dwight D. "Daily Log of Truck Train." Dwight D. Eisenhower Library, 1919

———. *At Ease: Stories I Tell to Friends*. Washington, DC: Eastern National, 1967.

———. *Mandate for Change, 1953–1956: The White House Years*. Garden City, NY: Doubleday, 1963.

Eisenhower, Susan. *How Ike Led: The Principles Behind Eisenhower's Biggest Decisions*. New York: Thomas Dunne Books, 2020.

Elkind, Sarah. *How Local Politics Shape Federal Policy: Business, Power, and the Environment in Twentieth-Century Los Angeles*. Chapel Hill: University of North Carolina Press, 2011.

Elwell-Sutton, L. P. *Persian Oil: A Study in Power Politics*. London: Laurence and Wishart, 1955.

Engler, Robert. *The Politics of Oil: A Study of Private Power and Democratic Directions*. New York: The Macmillan Company, 1961.

Feis, Herbert. *Petroleum and American Foreign Policy*. Stanford, CA: Food Research Institute, 1944.

Fischer, Louis. *Oil Imperialism: The International Struggle for Petroleum*. New York: International Publishers, 1926.

Flink, James J. *The Automobile Age*. Cambridge, MA: MIT Press, 1990.

Foley, Paul. "Petroleum Problems of the World War: Study in Practical Logistics." *United States Naval Institute Proceedings* 50 (1924): 1802–32.

Frankel, Paul. *Mattei: Oil and Power Politics*. New York and Washington: Praeger, 1966.

———. "Oil Supplies During the Suez Crisis: On Meeting a Political Emergency." *Journal of Industrial Economics* 6 (1958): 85–100.

Friedman, Thomas L. *From Beirut to Jerusalem*. New York: Farrar Straus Giroux, 1989.

Gartman, David. *Auto Opium: A Social History of American Automobile Design*. London: Routledge, 1994.

Gelbspan, Ross. *The Heat Is On: The Climate Crisis*. Reading, MA: Perseus Books, 1995.

Giddens, Paul. *Early Days of Oil: A Pictorial History of the Beginnings of the Industry in Pennsylvania*. Gloucester, MA: Peter Smith, 1964.

Goralski, Robert, and Russell W. Freeburg. *Oil and War: How the Deadly Struggle for Fuel in WWII Meant Victory or Defeat*. New York: William Morrow, 1987.

Gordon, Robert B., and Patrick M. Malone. *The Texture of Industry: An Archaeological View of the Industrialization of North America*. New York: Oxford University Press, 1994.

Gorman, Hugh. *Redefining Efficiency: Pollution Concerns, Regulatory Mechanisms, and Technological Change in the U.S. Petroleum Industry*. Akron, OH: University of Akron Press, 2001.

Greany, William C. "Daily Log of Truck Train." Dwight D. Eisenhower Library.

Greene, Ann. *Horses at Work: Harnessing Power in Industrial America* Cambridge, MA: Harvard University Press, 2009.

Gutfreund, Owen D. *Twentieth-Century Sprawl: Highways and the Reshaping of the American Landscape*. New York: Oxford University Press, 2005.

Hamilton, Adrian. *Oil: The Price of Power*. London: Michael Joseph/ Rainbird, 1986.

Hamilton, Shane. *Trucking Country: The Road to America's Wal-Mart Economy*. Princeton, NJ: Princeton University Press, 2008.

Hartshorn, J. E. *Oil Companies and Governments: An Account of the International Oil Industry in Its Political Environment.* London: Faber and Faber, 1962.

Heiss, Mary Ann. *Empire and Nationhood: The United States, Great Britain, and Iranian Oil, 1950–1954.* New York: Columbia University Press, 1997.

Hitchcock, William I. *The Age of Eisenhower: America and the World in the 1950s.* New York: Simon & Schuster, 2018.

Horowitz, Daniel. *Jimmy Carter and the Energy Crisis of the 1970s.* New York: St. Martin's, 2005.

Hughes, Thomas. *American Genesis.* New York: Penguin, 1989.

———. *Networks of Power: Electrification in Western Society, 1880–1930.* Baltimore: Johns Hopkins University Press, 1983.

Ickes, Harold L. *Fightin' Oil.* New York: Alfred A. Knopf, 1943.

Ise, John. *The United States Oil Policy.* New Haven, CT: Yale University Press, 1926.

Jackson, Elwell. "Daily Log of Truck Train." Dwight D. Eisenhower Library.

Jackson, Kenneth T. *Crabgrass Frontier: The Suburbanization of the United States.* New York: Oxford University Press, 1985.

Jentleson, Bruce. *Pipeline Politics: The Complex Political Economy of East-West Energy Trade.* Ithaca, NY: Cornell University Press, 1986.

Johnson, Robert. *Carbon Nation: Fossil Fuels in the Making of American Culture.* Lawrence: University of Kansas Press, 2014.

Jones, Christopher. *Routes of Power: Energy and Modern America.* Cambridge, MA: Harvard University Press, 2016.

Jones, Geoffrey, and Clive Trebilcock. "Russian Industry and British Business, 1910–1930: Oil and Armaments." *Journal of European Economic History* 11, no. 1 (1982): 61–104.

Jones, Toby Craig. *Desert Kingdom: How Oil and Water Forged Modern Saudi Arabia.* Cambridge, MA: Harvard University Press, 2010.

Juhasz. Antonia. *Tyranny of Oil: The World's Most Powerful Industry— and What We Must Do to Stop It.* New York: William Morrow, 2008.

Kane, N. Stephen. "Corporate Power and Foreign Policy: Efforts of American Oil Companies to Influence United States Relations with Mexico, 1921–28." *Diplomatic History* 1, no. 2 (1977): 170–98.

Kapstein, Ethan B. *The Insecure Alliance: Energy Crises and Western Politics Since 1944.* Oxford: Oxford University Press, 1990.

Kay, Jane Holtz. *Asphalt Nation: How the Automobile Took Over America and How We Can Take It Back.* Berkeley: University of California Press, 1997.

Kemp, Norman. *Abadan: A First-Hand Account of the Persian Oil Crisis.* London: Allan Wingate, 1953.

Kent, Marian. *Oil and Empire: British Policy and Mesopotamian Oil, 1900–1920.* London: Macmillan, 1976.

Kirsch, David. *The Electric Vehicle and the Burden of History.* Newark, NJ: Rutgers University Press, 2000.

Klare, Michael. *Blood and Oil: The Dangers and Consequences of America's Growing Dependency on Imported Petroleum.* New York: Holt, 2005.

———. *The Race for What's Left: The Global Scramble for the World's Last Resources.* New York: Picador, 2012.

———. *Resource Wars: The New Landscape of Global Conflict.* New York: Holt, 2002.

Klieman, Kairn. "From Kerosene to Avgas: International Oil Companies and Their Expansion in Sub-Saharan Africa, 1890s to 1945." In *Environment and Economics in Nigeria*, edited by Toyin Falola and Adam Paddock, 21–54. New York: Routledge, 2012.

———. "Oil, Politics, and Development in the Formation of a State: The Congolese Petroleum Wars, 1963–68." *International Journal of African Historical Studies* 41, no. 2 (2008): 169–202.

———. *The Pygmies Were Our Compass.* New York: Heineman, 2003.

———. "US Oil Companies, the Nigerian Civil War, and the Origins of Opacity in the Nigerian Oil Industry." *Journal of American History* 99, no.1 (2012): 155–65.

Koppes, Clayton R. "The Good Neighbor Policy and the Nationalization of Mexican Oil: A Reinterpretation." *Journal of American History* 69, no. 1 (1982): 62–81.

Lacy, Lee. "Dwight D. Eisenhower and the Birth of the Interstate Highway System," US Army. February 20, 2018. https://www.army.mil/article/198095.

Landau, Christopher T. "The Rise and Fall of Petro-Liberalism: United States Relations with Socialist Venezuela, 1945–1948." Senior Thesis, Harvard University, 1985.

Landes, David. *The Unbound Prometheus: Technological Change and Industrial Development in Europe.* New York: Cambridge University Press, 1969.

L'Espagnol de la Tramerye, Pierre. *The World Struggle for Oil.* Translated by C. Leonard Leese. London: George Allen & Unwin, 1924.

Levy, Walter J. *Oil Strategy and Politics, 1941–1981.* Edited by Melvin A. Conant. Boulder, CO: Westview, 1982.

Lewis, Tom. *Divided Highways: Building the Interstate Highways, Transforming American Life.* New York: Penguin Books, 1997.

Lieber, Robert J. *Oil and the Middle East War: Europe in the Energy Crisis.* Cambridge, MA: Harvard Center for International Affairs, 1976.

———. *The Oil Decade: Conflict and Cooperation in the West.* New York: Praeger, 1983.

Lifset, Robert. *American Energy Policy in the 1970s.* Norman: University of Oklahoma Press, 2014.

Lorant, Stefan. *Pittsburgh: The Story of an American City.* Lenox, MA: Authors' Editions, 1964.

Lovins, Amory. *Soft Energy Paths.* New York: Harper Collins, 1979.

Lubell, Harold. *Middle East Oil Crises and Western Europe's Energy Supplies.* Baltimore: Johns Hopkins University Press, 1963.

Maddow, Rachel. *Blowout: Corrupted Democracy, Rogue State Russia, and the Richest, Most Destructive Industry on Earth.* New York: Crown Publishing, 2019.

Malm, Andres. *Fossil Capital: The Rise of Steam Power and the Roots of Global Warming.* London: Verso Books, 2016.

———. How to Blow Up a Pipeline. London: Verso Books, 2021.

Marks, Robert. *The Origins of the Modern World: A Global and Environmental Narrative from the Fifteenth to the Twenty-First Century.* New York: Rowman & Littlefield, 2019.

Marriott, James. *The Oil Road: Journeys from the Caspian Sea to the City of London.* New York: Verso, 2013.

Marsh, Steve. *Anglo-American Relations and Cold War Oil: Crisis in Iran.* New York: Palgrave MacMillan, 2003.

Marx, Karl, and Friedrich Engels. *Manifesto of the Communist Party.* Chicago: Charles H. Kerr, 1906.

Mauch, Christof, and Thomas Zeller. *The World Beyond the Windshield: Roads and Landscapes in the United States and Europe.* Athens: Ohio University Press, 2008.

Mayer, Jane. *Dark Money: The Hidden History of the Billionaires Behind the Rise of the Radical Right.* New York: Anchor Books, 2017.

McCarthy, Joe. "The Lincoln Highway." *American Heritage* 25, no. 4 (1974): 32–37, 89.

McCarthy, Tom. *Auto Mania : Cars, Consumers, and the Environment.* New Haven, CT: Yale University Press, 2007.

McNaugher, Thomas L. *Arms and Oil: U . S . Military Strategy and the Persian Gulf.* Washington, DC: Brookings Institution, 1985.

———. "Walking Tightropes in the Gulf." In *The Iran-Iraq War: Impact and Implications*, edited by Efraim Karsh, 171-99. London: Macmillan, 1989.

McNeill, R. *Something New Under the Sun: An Environmental History of the Twentieth-Century World.* New York: Norton, 2001.

McNeill, R., and Peter Engelke. *The Great Acceleration: An Environmental History of the Anthropocene Since 1945.* Cambridge, MA: Harvard University Press, 2016.

McShane, Clay. *Down the Asphalt Path: The Automobile and the American City.* New York: Columbia University Press, 1994.

McShane, Clay, and Joel A. Tarr. *The Horse in the City: Living Machines in the Nineteenth Century.* Baltimore: Johns Hopkins University Press, 2008.

Mejcher, Helmut. *Imperial Quest for Oil: Iraq, 1910-1928.* London: Ithaca, 1976.

Melby, Eric D. K. *Oil and the International System: The Case of France, 1918-1969.* New York: Arno, 1981.

Melosi, Martin V. *Coping with Abundance: Energy and Environment in Industrial America.* New York: Knopf, 1985.

———. *Sanitary City.* Baltimore: Johns Hopkins University Press, 1999.

Melosi, Martin V., and Joseph A. Pratt. *Energy Metropolis: An Environmental History of Houston and the Gulf Coast.* Pittsburgh: University of Pittsburgh Press, 2008.

Meyer, Lorenzo. *Mexico and the United States in the Oil Controversy, 1917-1942.* Translated by Muriel Vasconcellos. 2nd ed. Austin: University of Texas Press, 1977.

Mikdashi, Zuhayr M., Sherrill Cleland, and Ian Seymour. *Continuity and Change in the World Oil Industry.* Beirut: Middle East Research and Publishing Center, 1970.

Miller, Aaron David. *Search for Security: Saudi Arabian Oil and American Foreign Policy, 1939-1949.* Chapel Hill: University of North Carolina Press, 1980.

Mitchell, Timothy. *Carbon Democracy: Political Power in the Age of Oil.* New York: Verso, 2013.

Mokyr, Joel. *Twenty-Five Centuries of Technological Change.* New York: Harwood Academic Publishers, 1990.

Moore, Frederick Lee, Jr. "Origin of American Oil Concessions in Bahrain, Kuwait, and Saudi Arabia." Senior Thesis. Princeton, NJ: Princeton University, 1948.

Mosley, Leonard. *Power Play: Oil in the Middle East*. New York: Random House, 1973.

Nemeth, Tammy. "Consolidating the Continental Drift: American Influence on Diefenbaker's National Oil Policy." *Journal of the Canadian Historical Association* 13, no. 1 (2002): 191–215.

——. "Continental Drift: Energy Policy and Canadian-American Relations." In *Diplomatic Departures: The Conservative Era in Canadian Foreign Policy, 1984–93*, edited by Nelson Michaud and Kim Richard Nossal, 59–70. Vancouver: UBC Press, 2001.

Nikiforuk, Andrew. *The Energy of Slaves: Oil and the New Servitude*. New York: Greystone, 2012.

——. *Tar Sands: Dirty Oil and the Future of a Continent*. New York: Greystone, 2010.

Norton, Peter D. *Fighting Traffic: The Dawn of the Motor Age in the American City*. Boston: MIT Press, 2008.

Nowell, Gregory Patrick. "Realpolitik vs. Transnational Rent-Seeking: French Mercantilism and the Development of the World Oil Cartel, 1860–1939." PhD diss. Boston: MIT Press, 1988.

Nye, David E. *Consuming Power: A Social History of American Energies*. Cambridge, MA: MIT Press, 1998.

——. *Electrifying America: Social Meanings of a New Technology, 1880–1940*. Boston: MIT Press, 1999.

——. *American Technological Sublime*. Boston: MIT Press, 1996.

O'Brien, Dennis J. "The Oil Crisis and the Foreign Policy of the Wilson Administration, 1917–1921." PhD diss. University of Missouri, 1974.

Odell, Peter R. *Oil and World Power: Background of the Oil Crisis*. 8th ed. New York: Viking Penguin, 1986.

Oliens, Roger M., and Dianna Davids. *Oil and Ideology: The American Oil Industry, 1859–1945*. Chapel Hill: University of North Carolina Press, 1999.

Painter, David S. "International Oil and National Security." *Daedalus* 120, no. 4 (1991): 183–206.

——. *Oil and the American Century: The Political Economy of US Foreign Oil Policy, 1941–1954*. Baltimore: Johns Hopkins University Press, 1986.

———. "Oil and the Marshall Plan." *Business History Review* 58, no. 3 (1984): 359–83.

———. "Oil and World Power." *Diplomatic History* 17, no. 1 (1993): 159–70.

Palmer, Michael A. *Guardians of the Gulf: A History of America's Expanding Role in the Persian Gulf, 1833–1992*. New York: Harper Collins, 1992.

Patterson, Matthew. "Car Culture and Global Environmental Politics." *Review of International Studies* 26, no. 2 (2000): 253–70.

Patton, Phil. *Bug: The Strange Mutations of the World's Most Famous Automobile*. New York: Simon & Schuster, 2002.

Paxson, Frederic L. "The Highway Movement, 1916–1935." American Historical Review 51, no. 2 (1946): 236–53.

Penrose, Edith T. *The Large International Firm in Developing Countries: The International Petroleum Industry*. London: George Allen & Unwin, 1968.

Penrose, Edith, and E. F. Penrose. *Iraq: International Relations and National Development*. London: Ernest Bern, 1978.

Philby, H. St. J. B. *Arabian Days: An Autobiography*. London: Robert Hale, 1948.

———. *Arabian Jubilee*. London: Robert Hale, 1952.

———. *Arabian Oil Ventures*. Washington, DC: Middle East Institute, 1964.

———. *Saudi Arabia*. London: Ernest Bern, 1955.

Philip, George. *The Political Economy of International Oil*. Edinburgh: Edinburgh University Press, 1994.

Pinkus, Karen. *Fuel: A Speculative Dictionary*. Minneapolis: University of Minnesota Press, 2016.

Pirani, Simon. *Burning Up: A Global History of Fossil Fuel Consumption*. New York: Pluto, 2016.

Plourde, André. "Canada's International Obligations in Energy and the Free Trade Agreement with the United States." *Journal of World Trade* 24, no. 5 (1990): 35–56.

Pratt, Joseph A., and William H. Hale. *Exxon: Transforming Energy, 1973–2005*. Austin: University of Texas Press, 2013.

Priest, Tyler. *The Offshore Imperative: Shell Oil's Search for Petroleum in Postwar America*. Dallas: Texas A&M University Press, 2009.

Prorokova-Konrad, Tatiana. *Transportation and the Culture of Climate Change: Accelerating Ride to Global Crisis.* Morgantown: West Virginia University Press, 2020.

Rabe, Stephen G. *The Road to OPEC: United States Relations with Venezuela, 1919–1976.* Austin: University of Texas Press, 1982.

Radocchia, Dino. *"Windows of Opportunity in Canadian-American Energy Relations."* Research Programme in Strategic Studies Occasional Paper No. 7. Downsview, Ontario: York University, 1987.

Rand, Christopher. *Making Democracy Safe for Oil: Oilmen and the Islamic East.* Boston: Little, Brown, 1975.

Randall, Stephen J. "Harold Ickes and United States Foreign Petroleum Policy Planning, 1939–1945." *Business History Review* 57, no. 3 (1983): 367–87.

——. "The International Corporation and American Foreign Policy: The United States and Colombian Petroleum, 1920–1940." *Canadian Journal of History* 9, no. 2 (1974): 179–96.

——. *United States Foreign Oil Policy, 1919–1948: For Profits and Security.* Kingston: McGill-Queen's University Press, 1985.

Rhodes, Richard. *Energy: A Human History.* New York: Simon & Schuster, 2018.

Rifkin, Jeremy. *The Hydrogen Economy.* New York: Penguin, 2003.

Risch, Erna. *Fuels for Global Conflict.* Washington, DC: Office of Quartermaster General, 1945.

——. *Quartermaster Support of the Army: A History of the Corps, 1775–1939.* Washington, DC: Center of Military History, 1989.

Rishel, Virginia. *Wheels to Adventure: Bill Rishel's Western Routes.* Salt Lake City: Howe Brothers, 1983.

Robertson, Thomas, Richard P. Tucker, Nicholas B. Breyfogle, and Peter Mansoor, eds. *Nature at War: American Environments and World War II.* New York: Cambridge University Press, 2020.

Rouhani, Fuad. *A History of OPEC.* New York: Praeger, 1971.

Rustow, Dankwart A. *Oil and Turmoil: America Faces OPEC and the Middle East.* New York: W. W. Norton, 1982.

Sabin, Paul. *The Bet: Paul Ehrlich, Julian Simon, and Our Gamble over Earth's Future.* New Haven, CT: Yale University Press, 2013.

——. *Crude Politics: The California Oil Market, 1900–1940.* Berkeley: University of California Press, 2005.

Salas, Miguel Tinker "Staying the Course: United States Oil Companies in Venezuela, 1945–1958." *Latin American Perspectives* 32, no. 2 (2005): 147–70.

Sampson, Anthony. *The Seven Sisters: The Great Oil Companies and the World They Shaped.* New York: Viking Adult, 1975.

Santiago, Myrna. *Ecology of Oil: Environment, Labor, and the Mexican Revolution, 1900–1938.* New York: Cambridge University Press, 2006.

Shaffer, Edward H. *Canada's Oil and the American Empire.* Edmonton: Hurtig, 1983.

Shulman, Peter. *Coal and Empire: The Birth of Energy Security in Industrial America.* Baltimore: Johns Hopkins University Press, 2019.

Shwadran, Benjamin. *The Middle East, Oil and the Great Powers.* 3rd ed. New York: John Wiley, 1973.

Skeet, Ian. *OPEC: Twenty-Five Years of Prices and Politics.* New York: Cambridge University Press, 1988.

Smil, Vaclav. *Energy and Civilization: A History.* Cambridge, MA: MIT Press, 2018.

———. *Energy in China's Modernization: Advances and Limitations.* Armonk, NY: M. E. Sharpe, 1988.

Smith, Jean Edward. *Eisenhower in War and Peace.* New York: Random House, 2012.

Snell, Mark A. *Gettysburg's Other Battle: The Ordeal of an American Shrine During the First World War.* Kent, OH: Kent State University Press, 2018.

Stent, Angela. *Soviet Energy and Western Europe.* Washington Papers 90. New York: Praeger, 1982.

Stivers, William. *Supremacy and Oil: Iraq, Turkey, and the Anglo-American World Order, 1918–1930.* Ithaca, NY: Cornell University Press, 1982.

Stoff, Michael B. *Oil, War, and American Security: The Search for a National Policy on Foreign Oil, 1941–1947.* New Haven, CT: Yale University Press, 1980.

Szeman, Imre. *On Petrocultures.* Morgantown: West Virginia University Press, 2018.

Szeman, Imre, and Jeff Diamanti. *Energy Culture.* Morgantown: West Virginia University Press, 2019.

Tarbell, Ida. *All in the Day's Work: An Autobiography.* Champaign: University of Illinois Press, 2003.

Tarr, Joel, ed. *Devastation and Renewal: An Environmental History of Pittsburgh and Its Region.* Pittsburgh: University of Pittsburgh Press, 2003.

———. *The Search for the Ultimate Sink: Urban Pollution in Historical Perspective.* Akron, OH: University of Akron Press, 1996.

Thynne, John F. "British Policy on Oil Resources, 1936–1951, with Particular Reference to the Defense of British Controlled Oil in Mexico, Venezuela and Persia." PhD diss. London School of Economics, 1987.

Trachtenberg, Alan. *Incorporation of America: Culture and Society in the Gilded Age.* New York: Hill and Wang, 1982.

Turner, Louis. *Oil Companies in the International System.* London: George Allen & Unwin, 1978.

Venn, Fiona. *The Oil Crisis.* London: Longman Pearson Education, 2002.

———. *Oil Diplomacy in the Twentieth Century.* New York: St. Martin's, 1986.

Vernon, Raymond. *Two Hungry Giants: The United States and Japan in the Quest for Oil and Ores.* Cambridge, MA: Harvard University Press, 1983.

———, ed. *The Oil Crisis in Perspective.* New York: W. W. Norton, 1976.

Vietor, Richard H. K. *Energy Policy in America Since 1945: A Study of Business-Government Relations.* New York: Cambridge University Press, 1984.

Vieyra, Daniel I. *Fill 'er Up: An Architectural History of America's Gas Stations.* New York: Macmillan, 1979.

Vollmann, William T. *Carbon Ideologies.* 2 vols. New York: Penguin, 2019.

Ward, Thomas E. *Negotiations for Oil Concessions in Bahrain, El Hasa (Saudi Arabia), the Neutral Zone, Qatar and Kuwait.* Privately printed. New York: 1965.

Weingroff, Richard F. "The Lincoln Highway." Federal Highway Administration. June 30, 2023. https://highways.dot.gov/highway-history/general-highway-history/lincoln-highway.

———. "The Man Who Changed America, Part I." Public Roads 66, no. 5 (March/April 2003). https://highways.dot.gov/public-roads/marchapril-2003/man-who-changed-america-part-i.

———. "Zero Milestone—Washington, DC." Federal Highway Administration. June 27, 2017. https://www.fhwa.dot.gov/infrastructure/zero.cfm.

Wells, Christopher. *Car Country: An Environmental History.* Seattle: University of Washington Press, 2013.

Whitaker, Jan. *Tea at the Blue Lantern Inn: A Social History of the Tea Room Craze in America.* New York: St. Martin's, 2015.

Wickman, John. "Ike and 'The Great Truck Train'—1919." Kansas History 13, no. 3 (1990): 139–48.

Williamson, Harold and Arnold R. Daum. *The American Petroleum Industry, Vol. I: The Age of Illumination 1859–1899.* Evanston, IL: Northwestern University Press, 1959.

Williamson, Harold F., Ralph L. Andreano, Arnold R. Daum, and Gilbert C. Klose. *The American Petroleum Industry, Vol. II: The Age of Energy, 1899–1959.* Evanston, IL: Northwestern University Press, 1959.

Yergin, Daniel. *The Prize: The Epic Quest for Oil, Money, and Power.* New York: Free Press, 1993.

———. *The Quest: Energy, Security, and the Remaking of the Modern World.* New York: Penguin, 2012.

Zallen, Jeremy. *American Lucifers: The Dark History of Artificial Light, 1750–1865.* Cambridge, MA: Harvard University Press, 2019.

Photos throughout the book courtesy of the National Archives.

Notes

1   Smith, *Eisenhower in War and Peace*, 19.
2   Ibid., 10.
3   Ibid., 8.
4   Eisenhower, *At Ease*, 37.
5   Smith, 18–19.
6   Eisenhower, *At Ease*, 104; Smith, 17–19.
7   Eisenhower, *At Ease*, 108.
8   General biographical information has been drawn from Smith, *Eisenhower in War and Peace*.
9   Snell, *Gettysburg's Other Battle*, 70–71.
10 Marx, *Communist Manifesto*, 17.
11 Ambrose, *Eisenhower: Soldier and President*, 28.
12 Ibid., 30.
13 *New York Sun*, 1898. Quoted in Kirsch 2000, 11.
14 McCarthy, *Auto Mania*, 34–37.
15 Ibid., 38–39.
16 Smith, 38–40.
17 Snell, 142–47.
18 Eisenhower, *At Ease*, 42–50.
19 Ibid., 39–43.
20 Black, *Crude Reality,* 127–129.
21 Snell, 157.
22 Ibid.
23 Eisenhower, 136.
24 Ibid., 136; Snell, 149.
25 Eisenhower, *At Ease*, 137.
26 Snell, 77.

27   Ibid., 170.
28   Ibid., 149.
29   Ibid., 173.
30   Ibid.
31   Ibid., 176–8.
32   Snell contains a general description of these training efforts.
33   Ibid., 174–7.
34   Ibid., 177–8.
35   Eisenhower, *At Ease*, 169–170.
36   Ibid., 146; Snell, 160.
37   Ibid., 158–60.
38   Eisenhower, *At Ease*, 147.
39   Snell, 200.
40   Eisenhower, *At Ease*, 148.
41   Snell, 204.
42   Ibid., 213–214; Smith, 47.
43   Eisenhower, *At Ease*, 155–6; Smith, 48.
44   Eisenhower, *At Ease*, 155.
45   Ibid., 157.
46   Davies, *American Road*, 8.
47   Eisenhower, *At Ease*, 151–153.
48   E. R. Jackson's Report to L. B. Moody, October 31, 1919.
49   General details are drawn from Davies, 7–9.
50   *Washington Evening Star*, July 7, 1919. This coverage was then picked up by other newspapers nationally.
51   Ibid.
52   Ibid.
53   Weingroff, "Zero Milestone."
54   Eisenhower, *At Ease*, 157.
55   Eisenhower, "Daily Log of Truck Train."
56   Eisenhower, *At Ease*, 158.
57   Davies, 44.
58   Ibid., 19–21.
59   Ibid., 23.
60   Ibid., 24.
61   Yergin, *The Prize*, 80.
62   Kay, *Asphalt Nation*, 154–158.
63   For more information on this topic, see Black, *Crude Reality*, 78–80.
64   Black, *Internal Combustion*, 130–131.

65   Ibid., 136, 140, 148.

66   *Wall Street Journal,* May 27, 1914, from Black, *Internal Combustion,* 156.

67   For more information on this topic, see Black, *Crude Reality,* 79–81.

68   Black, *Crude Reality,* 119.

69   Davies, 25–26.

70   Ibid., 31–34.

71   Ibid., 35–37.

72   For more information on this topic, see Black, *Crude Reality,* 78–80.

73   Whitaker, *Tea at the Blue Lantern,* 18.

74   "At the Sign of the Tea-Room," *Good Housekeeping,* July 1917, quoted in Whitaker, 59, 62.

75   McCarthy, *Auto Mania,* 47.

76   Vieyra, *Fill 'er Up,* 1–5.

77   For more information on this topic, see Black, *Crude Reality,* 97–98; 119–121.

78   Vieyra, 7.

79   For more discussion of this topic, see Black, *Crude Reality,* 98.

80   Black, *Crude Reality,* 120.

81   Davies, 32–35.

82   Ibid., 47–48. Davies reports that Johnson delivered some variation of this speech almost every night of the trip and that it was dispersed widely in numerous newspapers. Eisenhower, *At Ease,* 159.

83   Jackson's Report to Moody, Oct. 31, 1919.

84   Ibid. A copy of the film can be accessed at the Eisenhower Library.

85   Ibid.

86   Ibid.

87   Davies, 50.

88   Lorant, *Pittsburgh.* Quoted in Davies, 51–52.

89   Personal Report filed from Eisenhower to Chief Motor Transport Corps, Nov. 3, 1919.

90   The details of the convoy are drawn from a few specific sources: The Daily Log kept by Elwell Jackson, Captain Greany's Daily Report, and collected telegrams from Colonel McClure. Capturing these and other sources as well, Davies, *American Road* stands as the most complete source for actual events on the trek.

91   Davies, 56–58.

92   Ibid., 67.

93   Ibid., 70, 67.

94   Ibid., 69–70, 74.

95   Ibid., 77–78.
96   Eisenhower to CMTC.
97   Ibid.
98   Jackson to Moody.
99   Eisenhower to CMTC.
100   Jackson to Moody.
101   See Rebecca Conard, "The Lincoln Highway in Greene County."
102   Davies, 83–86.
103   *Cleveland Plains Dealer*, July 27, 1919.
104   Davies, 84–87.
105   Ibid., 16–18.
106   Ibid., 11–17, 25–26.
107   Ibid., 26–27.
108   Ibid., 30; Weingroff, "The Lincoln Highway."
109   Ibid.
110   Davies, 30–32.
111   Weingroff, "The Lincoln Highway."
112   Ibid.
113   Davies, 34.
114   Ibid., 34–6.
115   Risch, *Quartermaster Support of the Army*, 597; Davies, 35–37.
116   Ibid., 36–38.
117   Ibid., 80–83.
118   Ibid.
119   *Colliers*, Nov. 23, 1916. Quoted in Davies, 83.
120   "History of Clinton County, Iowa," Clinton County Historical Society, 1976; Davies, 82–85.
121   Davies, 82–85.
122   Ibid.
123   Ibid., 58–59.
124   Ibid.
125   Ibid., 94–5.
126   *Davenport Democrat*, quoted in Davies, 88; Davies, 91.
127   Ibid., 94–95.
128   Ibid., 96–97.
129   Ibid., 97–98.
130   Ibid., 102.
131   Ibid., 115.
132   Ibid., 115–116, 120–124.

133 Ibid., 125–126.

134 Ibid.

135 Eisenhower, *At Ease*, 161.

136 Ibid., 157.

137 Ibid., 160.

138 Ibid., 162–163.

139 Ibid., 162–164.

140 Ibid., 164.

141 Ibid., 165.

142 Ibid., 158.

143 Davies, 94–95.

144 Eisenhower, *At Ease*, 161.

145 The details of the convoy are drawn from a few specific sources: The Daily Log kept by Elwell Jackson, Captain Greany's Daily Report, and collected telegrams from Colonel McClure. Capturing these and other sources as well, Davies, *American Road* stands as the most complete source for actual events on the trek.

146 Black, *Crude Reality*, 140–142; Santiago, *Ecology of Oil*, 156.

147 Winston Churchill, House of Commons, June 17, 1914.

148 Ibid.

149 Davies, 208.

150 Ibid., 210–200.

151 Ibid., 208.

152 Davies, 211. *San Jose Mercury News* (published as *San Jose Mercury Herald*), Sep. 7, 1919.

153 Davies, 133.

154 Ibid., 141.

155 Ibid., 145.

156 Ibid.

157 Ibid., 164–166.

158 Ibid., 165–166.

159 Ibid., 161–168.

160 Rishel, *Wheels to Adventure*.

161 Ibid.; Davies, 171.

162 Davies, 172–175.

163 Ibid.

164 Ibid. 174–177.

165 Ibid.

166 Jackson to Moody.

167 Davies, 176.
168 Rishel, *Wheels to Adventure*; Davies, 176.
169 Jackson to Moody.
170 Davies, 182–183.
171 Ibid., 184, 190.
172 Ibid., 191–194.
173 Ibid., 187; Jackson's Daily Log.
174 Ibid.
175 Ibid.; Davies, 201.
176 Ibid., 201–204.
177 Ibid.; Kevin Cook, "Ike's Road Trip," HistoryNet.com, March 26, 2019
178 Davies, 206.
179 Ibid., 206–211.
180 Davies, 211.
181 Kevin Cook, "Ike's Road Trip," HistoryNet.com, March 26, 2019.
182 Davies, 211.
183 Ibid., 213–216.
184 Ibid.
185 Ibid., 221–223.
186 Ibid., 222–225; 228–230.
187 Ibid., 230–231.
188 Eisenhower, *At Ease*, 166–167.
189 Ambrose, 44.
190 Smith, 148–152.
191 Ambrose, 52.
192 Ibid., 60–61.
193 Ibid., 61.
194 Smith, 362–364.
195 Ibid., 367–368.
196 Ibid., 459.
197 For more information on this topic, see Black, *Crude Reality*.
198 Gartman, *Auto Opium*, 157.
199 For more information on this topic, see Black, "Oil for Living."
200 Gartman, 149.
201 Ibid., 174–177.
202 For more information on this topic, see Black, "Oil for Living."
203 Gartman, 141.
204 Lewis, *Divided Highways*, 84.

205 Cohen, *Consumer's Republic*, 70, 119.
206 McShane, *Down the Asphalt Path.*
207 For more information on this topic, see Black, *Crude Reality*, 158–159.
208 Jackson, *Crabgrass Frontier*, 163.
209 Ibid., 164. Jackson reports that a variety of surfaces were in use for early roads, including crushed stone tied together called macadam; asphalt, which was similar but relied on underpinnings; and concrete. The first two materials relied on petroleum for their production.
210 Ibid., 167, 170.
211 Ibid., 170–171.
212 See Black, *Climate Change.*
213 For more information on this topic, see Black, *Crude Reality*, 120–122; Black, "Oil for Living."
214 Patton, *Bug.*
215 Weingroff, "The Man Who Changed America."
216 Eisenhower, *Mandate for Change*; Weingroff, "Zero Milestone"; Eisenhower, *At Ease*, 166–7.
217 Hitchcock, *The Age of Eisenhower*, 245.
218 Ibid.
219 Ibid., 253.
220 Ibid., 254, 189.
221 Lewis, 98.
222 Hitchcock, 263.
223 Lewis, 106–107.
224 Lacy, "Dwight D. Eisenhower and the Birth of the Interstate Highway System." Overall discussion of the Clay Committee Report and full-text access can be found at https://www.fhwa.dot.gov/infrastructure/clay.cfm.
225 Lewis, 109.
226 Ibid., 103.
227 Lacy.
228 Ibid.
229 Ibid.
230 For a fuller discussion of each of these funding options, see Lewis, 113–118.
231 Ibid., 114–116.
232 Ibid., 119–22.
233 Ibid., 144–146.
234 McNeill and Engelke, *The Great Acceleration,* 9–11.
235 Ibid., 40.

Acknowledgments

This project's origins and creative cocoon have been very slight, which has made it particularly magical for me.

One scholar, who was not particularly persistent, one day let slide, "Why not?" Ted Widmer had contacted me to solicit an essay on the year 1919. I worked with Ted and the editors to create a brief 2019 piece that appeared in the *New York Times*. I spoke about it with Ted at longer length a few more times and at the end of each conversation, he again said, "Why not?" I thought I had many reasons to not, and, so I never followed up. COVID changed life for at least a year and at one point I too said, "Why not?"

Josh Bodwell at Godine inspired me further and I set out on my own trek to see if the story could fit together neatly. He presented me with a welcoming writing process that I did not think existed in the world. The major part of that process was Celia Johnson, my editor with whom I shared a creative year that, once again, I did not think existed. I thank each of them and Daniel for their support, interest, and effort. Without it, I truly would have moved on to other things and Ike would have remained as only my prop in teaching and lectures. This book exists because of their creative environment and the priorities of Godine.

I am similar to many scholars who for years of work with students flesh out relevant historical junctures that help current developments to make sense. Ike's road trip was one of those props in my courses on history and environmental studies at Penn State Altoona. My supportive environment there allowed me to take on this somewhat different writing project and to explore working in narrative history writing. I am thankful to Lori, Ken, Peter, Todd, Ed, Lisa, Nicholas, Carolyn, Erin, and Ian.

But, finally, our historic moment evolved in the early 21st century and made the road trip and energy transitions relevant again. They made this story particularly important to tell again so that it might be recast with an eye to our present shift in energy use. Truly, the story's time had re-emerged and I was in a place to tell it. More than ever, I felt a vehicle for work that needed to be out there.

None of my writing occurs without life's support structure. To Chrissy, Sam, Ben, Marta, Clyde, Jennifer, Owen and others go my undying appreciation for the opportunity to ruminate, conjure, and write. To Don, John, Adam, Paul, Neil, and others goes my appreciation for constructing the line of questions and interests that make me go. And, in the area of energy history inquiry, I express my gratitude to the many scholars who now contribute to our important work to be part of making our transition an informed, intelligent one.

"Why not?"

The Shack
Wellfleet, MA
February 2024

About the Author

Dr. Brian C. Black is Distinguished Professor of History and Environmental Studies at Penn State Altoona, where he also served for over a decade as Head of Arts and Humanities. Recognized as a global expert on energy and petroleum history, he is the author of more than a dozen books, which include: *Petrolia: The Landscape of America's First Oil Boom; Crude Reality: Petroleum in World History;* and *To Have and Have Not: Energy in World History.* His writing on energy has appeared in the *Christian Science Monitor, USA TODAY, the Conversation,* and the *New York Times.* He divides his time between central Pennsylvania and Cape Cod, Massachusetts.

A NOTE ON THE TYPE

Ike's Road Trip has been set in Bulmer Monotype. It is a spirited and transitional typeface which dates back to 1790 and is modeled upon Baskerville. Bulmer was designed and cut by William Martin for William Bulmer of the Shakespeare Printing office.

Book Design by Brooke Koven
Composition by Leslie Anne Feagley

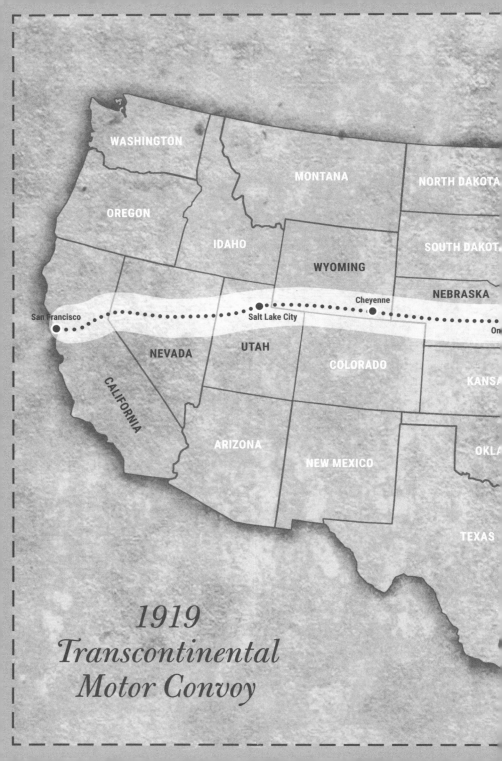

1919
Transcontinental
Motor Convoy